Practical Astronomy

Springer
London
Berlin
Heidelberg
New York
Barcelona
Budapest
Hong Kong
Milan
Paris
Santa Clara
Singapore
Tokyo

Other titles in this series

Seeing Stars

The Night Sky Through Small Telescopes

Chris Kitchin and Robert W. Forrest

Springer

Chris Kitchin, BA, BSc, PhD, FRAS
Robert W. Forrest, BSc, MSc, FRAS, LLCM

University of Hertfordshire, College Lane, Hatfield,
Hertfordshire AL10 9AB, UK

Cover illustration: The suburban night sky, showing Scorpius. (Photograph by Gerald North.)

ISBN 3–540–76030–X Springer-Verlag Berlin Heidelberg New York

British Library Cataloguing in Publication Data
Kitchin, Christopher R. (Christopher Robert), 1947–
   Seeing stars : the night sky through small telescopes. –
(Practical astronomy)
   1.Telescopes 2.Astronomy – Observations
   I.Title II.Forrest, Robert W.
   522.2
ISBN 354076030X

Library of Congress Cataloging-in-Publication Data
Kitchin, C. R. (Christopher R.)
   Seeing stars : the night sky through small telescopes / Chris
Kitchin and Robert W. Forrest.
      p.   cm. – (Practical astronomy)
   Includes index.
   ISBN 3–540–76030–X (paperback : alk. paper)
   1. Astronomy – Amateurs' manuals.   2. Astronomy – Observers'
manuals.   I. Forrest, Robert W., 1950–   . II. Title.
III. Series.
QB63.K48   1997   97–34693
522′.2–dc21   CIP

Typeset by EXPO Holdings, Malaysia
Printed and bound by Interprint Limited, Malta
58/3830-543210 Printed on acid-free paper

Preface

This book is intended for astronomers of any age and depth of theoretical knowledge who are just embarking on their first steps in observational astronomy. Its primary aim is to show the fascinating objects and phenomena that you can expect to see through a telescope and with the naked eye, and to suggest projects and observing programmes suitable for the beginner and the more experienced observer.

The primary emphasis, however, is on what you can EXPECT TO SEE. The images in most astronomy books are the best that can possibly be obtained. They come from using large telescopes, long exposures, highly sensitive detectors, and often have been subjected to sophisticated computer processing after the images have been obtained. The resulting glorious technicolour images of distant galaxies and nebulae are breathtaking, but they are not what you will *see* if you look through a telescope. This is true of any size of telescope but most especially it is true of the relatively small instruments likely to be used by apprentice observers. The majority of the images used in this book have therefore been reproduced to show what you may expect to see under normal conditions with the naked eye or through a telescope of 0.3 m (12 inches) aperture or less. Except where otherwise indicated, they have all been obtained by the authors using small telescopes or ordinary camera lenses.

We hope that if you read through this book before starting observing then you will find it much easier thereafter to find and observe the various objects for which you may be searching, because your expectations will be realistic. You will not be disappointed when a distant galaxy appears as a faint fuzz, and not a majestic spiral cartwheeling its way across the sky. The images obtained on large telescopes have their place in astronomy, or there would be no point in obtaining them in the first place, but so also do the observations made by the millions of enthusiastic part-time astronomers. Astronomy is unique among the sciences because of its enormous popular following (how many amateur solid-state physicists have you met?), and observations contribute greatly to that popularity. The satisfaction that you will obtain when you finally catch the first glimpse of a new nebula or comet, or show the spectacular mountains on the Moon to a young friend will soon far outweigh even the best images obtainable from the largest telescopes.

We wish you all clear skies and good observing!

Chris Kitchin and Robert Forrest
Hertford, 1997

Acknowledgements

The authors wish to thank John Watson for the original idea for this book. They would also like to thank Jeremy Bailey for obtaining the images of the southern constellations and Paul Martin for printing innumerable apparently totally blank negatives with unfailing good humour.

Contents

Chapter 1

Finding Your Way Around the Sky

1.1 Introduction

Buying the telescope was the easy part. Now however you are wondering how to cope with the apparent complexities of finding, in the sky, the innumerable objects that you have seen in the astronomy magazines and books. Do not despair: it is a lot easier than you might think.

If you have not yet bought your telescope, then it would be a good idea to read Chapter 2 and especially to look at Table 2.1 first. If the telescope has direct read-outs of the position that it is pointing to in the sky (called Right Ascension and Declination – Section 1.4), or is computer controlled, then you may want to try setting on things immediately. BUT YOU DO NOT HAVE TO HAVE THESE EXTRAS. Most objects that you will be able to see in your telescope, even those that are quite faint, can be found directly. It helps to have a good finder telescope attached to your main telescope (Section 2.8.4), but even this is not essential.

You do not need to know the precise position that your telescope is pointing to (or even to know it at all) in order to find a new object, because there are so many objects in the sky that, once you can recognise a few of them, you can soon move from the known ones to the unknown. This process is called *star hopping* and it is discussed in more detail in Sections 1.3 and 2.8.4. The first step, however, is to begin to be able to recognise the patterns to be seen in the sky with the naked eye; in other words, to be able to recognise the constellations.

1.2 Constellations

Probably, if you have started to read this book, then you can already recognise some of the constellations. If not, you may well know someone who can point some of them out to you; boy scouts and girl guides are useful here. Alternatively you could go along to your local astronomical society (Appendix 1) where you will undoubtedly find yourself deluged with help.

Let us however start from scratch and assume that you need to begin finding your way around the sky without help. There are 88 officially recognised constellations, of which 50 or 60 will be visible at some time or other from the average observing site. Learning the constellations therefore sounds a very formidable task. Fortunately you do not need to know them all. A dozen or so of the larger, brighter constellations will be enough to enable you to find your way around the sky with complete confidence. Moreover, even those constellations may be learnt a few at a time, as the gradual progression of the seasons changes our view of the heavens. Even knowing just one constellation may be enough; provided, of course, that it is the one that contains the object for which you are searching. The major constellations will be more than sufficient to find your way around the sky at all times of night and of the year. Those listed in Table 1.1 will serve northern hemisphere observers, while those in Table 1.2 will enable southern observers to find their way around the sky.

The rest of this section is concerned with the aim of becoming familiar with the patterns in the sky. But before

Table 1.1. The major constellations for northern hemisphere observers

Andromeda	Aquila	Auriga	Boötes
Canis Major	Cassiopeia	Cygnus	Gemini
Leo	Lyra	Orion	Pegasus
Perseus	Taurus	Ursa Major	Ursa Minor

Table 1.2. The major constellations for southern hemisphere observers

Aquila	Ara	Canis Major	Carina
Centaurus	Cetus	Crux	Gemini
Grus	Leo	Lupus	Orion
Pavo	Pisces Austrinus	Puppis	Sagittarius
Scorpius	Taurus	Triangulum Australe	Vela

we proceed, there is one aspect of the task in which this book, and almost every other similar publication, may be misleading and therefore needs to be mentioned. This problem relates to the scale of things in the sky. The photographs in this section were obtained using a 35 mm camera with a 17 mm lens. They therefore cover about a quarter of the hemisphere of the sky that you may be seeing. Looking at the photographs, or using other star maps etc., can lead you to expect the constellations to be quite small. They are not; the major constellations cover tens of degrees of the sky. To get an idea of the sizes that you should expect, first of all use the scale provided with the photographs to estimate the height and width of the constellation. Then use the useful guide that, for most people, the clenched fist held at arm's length (see Fig. 1.1) covers about 8°, to step out the amount of sky that you should expect the constellation to cover. You will then at least start with the correct expectation of the size of the thing you are trying to find.

1.2.1 Getting Started

In the northern hemisphere, the best constellation to start with is the Plough, also known as the Great Bear, Ursa Major, King Charles' Wain, the Big Dipper, or less officially as the Saucepan (strictly, the Plough is formed from just the seven main stars of Ursa Major). This is because for latitudes to the north of about 40°N, the main stars of the constellation will always be visible on a clear night (that is, the stars are circumpolar – they never set). Even at the equator, some or all of the main stars of the Plough will be in the sky for 60 per cent of the time. It is

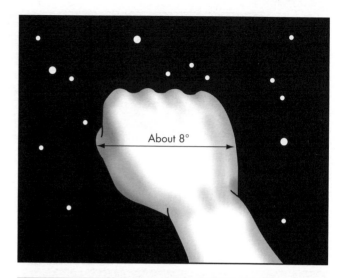

Figure 1.1 The clenched fist at arm's length spans about 8°.

often a puzzle why constellations have the names that they do, because the shapes outlined by the main stars rarely resemble the object they are supposed to represent. In many cases, of course, the name has no association with shape, but this is not the case with the Plough, though we have to imagine the ancient horse-drawn implement (see Fig. 1.2) rather than the modern multi-tiller dragged by a tractor.

For observers in the southern hemisphere, Crux (the Southern Cross – see Figs 1.20, 1.22 and 1.23) is probably the best choice to start with, while for observers near the equator, Orion (the Hunter – see Figs 1.16, 1.17 and 1.18) is a good choice, provided that it is in the sky at the time,

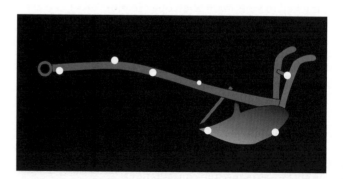

Figure 1.2 The plough (Ursa Major).

and not below the horizon. Both these constellations are easily recognisable *ab initio*, and Crux is completely circumpolar for observers south of 35°S. The procedure for finding Ursa Major is gone though in detail below. If you should be beginning with Orion or Crux, or indeed any other constellation, then a similar process will soon get you started.

How then do you find Ursa Major for the first time? One of the problems with finding any constellation that you have never recognised before is that its appearance will change depending on circumstances. Thus from an urban observing site, probably only the seven main stars of the constellation will be visible (see Fig. 1.3 and Fig. 1.4, *overleaf*). In poor, hazy conditions and with a bright Moon, you may not be able to see any stars at all. From a reasonable site, on a good clear night, without a Moon on the other hand, you will probably be able to see two or three dozen stars in the constellation (see Fig. 1.5, *overleaf*). The appearance will also depend on the orientation of the constellation (see Figs 1.6 and 1.20, page 13). This is less variable for constellations near the equator, but for those near the poles, like Ursa Major and Crux, they can be at any angle. Finally, the constellation may be at different altitudes above the horizon; or from some latitudes, some or all the stars may be below the horizon. For Ursa Major, the constellation will have the appearance of Fig. 1.6c when it is highest in the sky, and that of Fig. 1.6a when it is lowest. Figure 1.6d shows it at its western-most position, and Fig. 1.6b at its eastern-most position.

With these problems in mind therefore, it is best to select a reasonably clear moonless night and, if need be, to move away from town and other bright lights, for your first attempt at constellation identification. As you are

---

Magnitude Scale on the Star Maps

● Magnitude 1.5 and brighter

● Magnitude 2.5 to 1.5

• Magnitude 3.5 to 2.5

· Magnitude 4.5 to 3.5

· Magnitude 4.5 and fainter

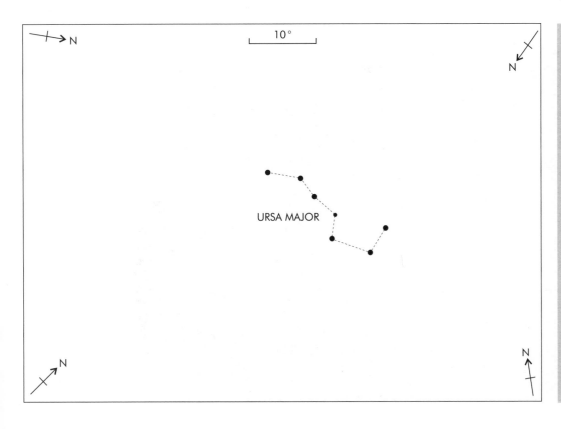

URSA MAJOR

Figure 1.3 The seven major stars of Ursa Major – also called the Plough. (Note: The lines joining the stars on this and subsequent figures are only there to help fix the pattern of the constellation in your mind. They have no other significance.) All photographs and their accompanying star maps in this section are shown at exactly the same scale and can therefore be compared easily.

Figure 1.4 The Ursa Major region – stars visible from a typical urban, light-polluted site.

Figure 1.5 The Ursa Major region – stars visible from a good site.

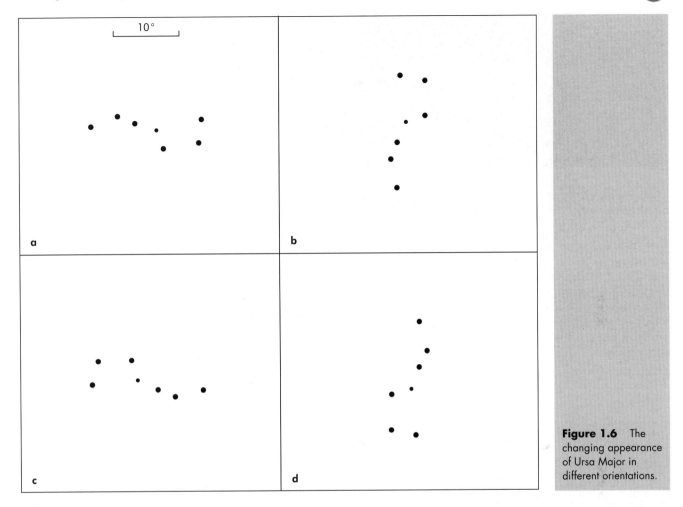

Figure 1.6 The changing appearance of Ursa Major in different orientations.

able to recognise more and more constellations, you will find that it is not essential to have good observing conditions in order to find your way around the sky, but it will help considerably to begin with. You will also need to know the approximate direction of north on the ground. If you do not know this already for your area, then it can easily be obtained from a good-quality map. Many will have a compass rose marked, but if not, maps in the northern hemisphere are almost invariably oriented with north at the top.

Having chosen a reasonable night, arrived at a reasonable observing site, and oriented yourself to face north, what then? First get an idea of the area of the sky over which you will need to search. Ursa Major is about 35° from the North Pole. You will therefore need to search over a circle extending from 35° either side of north, and from near the horizon to near the zenith. This radius is

about four times the width of the clenched fist at arm's length (see Fig. 1.1). The constellation's seven main stars extend over a distance of about 20° (2.5 × the clenched fist at arm's length). Step these two distances out so that you can visualise the scale of the constellation and of the search area. Then with the possible appearances in mind (see Fig. 1.6), look among the brighter stars in the search area until you can find the saucepan-like pattern of the Plough (see Fig. 1.7, *overleaf*). You should be able to spot the constellation within a minute or two.

If after ten or fifteen minutes' searching you have not found the constellation, then you are probably not facing north, so check your orientation. Alternatively, if your latitude is south of about 40°N, then some or all of the stars of Ursa Major may be below your horizon (astronomy magazines and some national newspapers publish monthly sky charts showing the stars visible for the appropriate

6

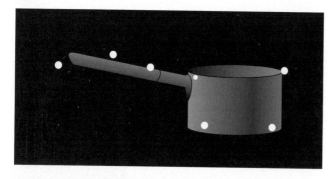

Figure 1.7 Ursa Major as a Dipper or Saucepan.

Assuming that you have been successful in finding the main stars of Ursa Major, you can then start to become familiar with their individual names. Many stars, especially the brighter ones, can have several names. Thus the seven bright stars of Ursa Major are known as Dubhe, Merak, Phad, Megrez, Alioth, Mizar and Alkaid (see Fig. 1.8). Mizar has a fainter close companion called Alcor that can be seen next to it if you have average or better eyesight. The seven main stars also have designations under the Bayer system as α Ursa Majoris (usually abbreviated as α UMa – see Appendix 6 for the genitives and abbreviations of all the constellation names), β UMa, γ UMa, δ UMa, ε UMa, ζ UMa, and η UMa (see Fig. 1.9). Under the Bayer system, the brighter stars within a constellation are given Greek-letter labels. Usually, this is in order of their brightnesses: so that α designates the brightest star of the constellation, β the second brightest and so on (the full Greek alphabet is listed in Appendix 5). However since you can clearly see (see Figs 1.4 and 1.5) that δ UMa (Megrez) is fainter than ε UMa, ζ UMa and η UMa (Alioth, Mizar and Alkaid), the system has its glitches. Stars may also be known by their numbers in various star catalogues. Thus Dubhe (α UMa) is also

time of year, so you can check these if you suspect that Ursa Major is below the horizon). In the latter case you can either try again at a different time of night or later in the year, or start with a different constellation. If Ursa Major should be visible and you are oriented correctly to see it, and yet you still cannot find it, then you will probably have to find someone who can point it out to you directly.

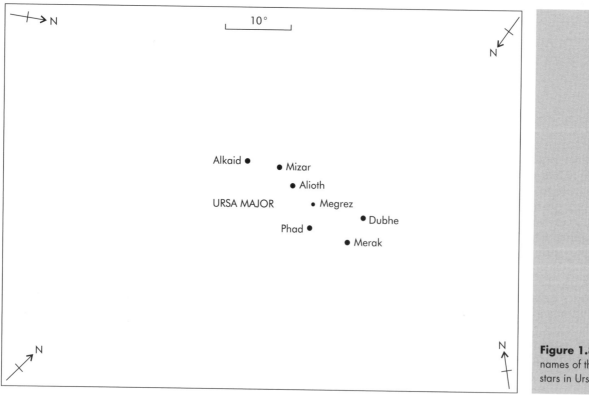

Figure 1.8 The names of the main stars in Ursa Major.

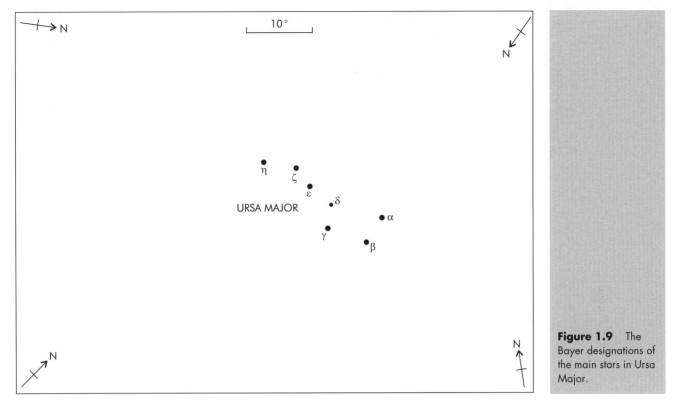

Figure 1.9 The Bayer designations of the main stars in Ursa Major.

called HD 95689 (from the Henry Draper catalogue), BS4301 (from the Bright Star catalogue), BD + 62° 1161 (from the Bonner Durchmusterung catalogue), or as 50 UMa (from the Flamsteed catalogue). For purposes of recognising the constellations, however, the star's name or Bayer designation will generally be sufficient.

Once you are familiar with the main stars of Ursa Major, you should start looking for the fainter ones (see Fig. 1.5 and Figs 1.10 and 1.12, *overleaf*). At this stage you may find a good star atlas (Appendix 2) useful as an addition to the maps and images given in this book.

As well as the naked-eye stars, each constellation is now used to delimit a specific area of the sky. These areas cover the whole sky and are officially recognised by the International Astronomical Union (see, for example, the outer boundary of Ursa Major marked in Fig. 1.10, *overleaf*). The boundaries were chosen so that they ran along lines of Right Ascension and Declination (Section 1.4) in 1875. Nowadays, precession (Section 1.4) has removed that alignment, though the boundaries have not moved with respect to the stars. Since the boundaries were designated in 1922, and many stars were named before that date, there are numerous anomalies in the system. Thus within the official boundaries of Ursa Major may be found the stars 1 CVn (Canes Venatici), 55 Cam (Camelopardalis) and 15 LMi (Leo Minor); while 10 UMa is actually in the area assigned to the constellation Lynx, etc. Fortunately, both the official boundaries and these latter complications are something you can completely ignore while learning your way around the sky.

1.2.2 Moving Onwards

Once you can identify Ursa Major (or any other "starter" constellation), finding other constellations becomes much easier. Ursa Major is a good starter constellation because it provides helpful guides to its surrounding constellations. The end stars (α and β UMa) indeed are often called the *pointers*, because they may be used to find the Pole Star (α Ursae Minoris, Polaris – Fig. 1.11, *overleaf*). The Pole Star is a reasonably bright star that by chance happens to be very close to the position of the North Pole in the sky. It is therefore very useful as a guide to finding your orientation on the surface of the Earth. To find Polaris simply follow the line of the pointers for about five times their

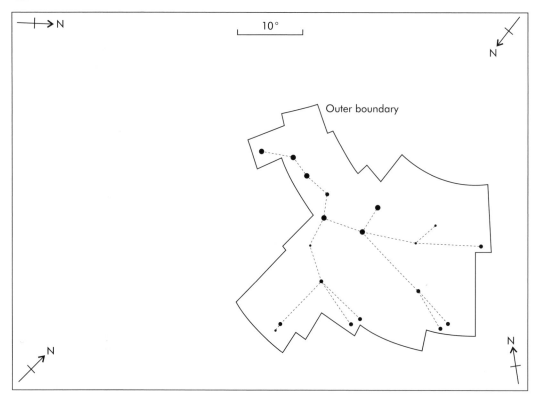

Figure 1.10 Ursa Major showing the stars usually visible to the naked eye from a good site, plus its official outer boundary.

separation, and you come directly to the star (see Fig. 1.11). Polaris is the brightest star in the constellation Ursa Minor (Little Bear), and so the rest of that constellation can be easily identified once Polaris has been recognised (see Figs 1.5, 1.11 and Fig. 1.13 (*overleaf*)).

Orion (the Hunter – Fig. 1.15, *overleaf*) is another good starter constellation, and strongly resembles the figure it is supposed to represent (see Fig. 1.16, *overleaf*). For observers in equatorial latitudes, for whom Ursa Major and Crux (see below) are not circumpolar, Orion is probably the best constellation with which to commence constellation identification, although it is not visible every night. The stars of Orion's belt then provide the direction of the first star hop westwards to the brightest star in the night sky, Sirius (α CMa), and so to the rest of the constellation of Canis Major.

For observers south of the equator, the Southern Cross (Crux – Figs 1.19 and 1.21, *overleaf*) is a good starting point for constellation identification. Its main stars are circumpolar for any latitudes south of 35°S, and it is easily recognisable *ab initio*. It therefore forms the equivalent of Ursa Major as a starting point for most southern observers. Like Ursa Major, the appearance of Crux in the

sky changes with its orientation, but it is a compact, bright constellation and usually easily recognisable in any orientation (see Fig. 1.20, *overleaf*). Once identified, Crux provides a good guide to its neighbouring constellations Centaurus and Carina and to the southern pole, though in the south there is no convenient bright star to mark the pole.

1.3 Star Hopping

The idea of star hopping should be quite familiar now if you have worked through Section 1.2. The process is, however, not limited to finding new constellations. It may be used on any scale (see Figs 1.24, *overleaf* and 2.22), and in particular when looking through your telescope (see Sections 2.8.4 and 2.10). Once you know the constellations, you will often be able to find quite faint objects, just by estimating their position with respect to the stars that you can see, and then pointing your telescope by eye towards that position. You should use the lowest power (longest focal length) eyepiece available in order to have the widest

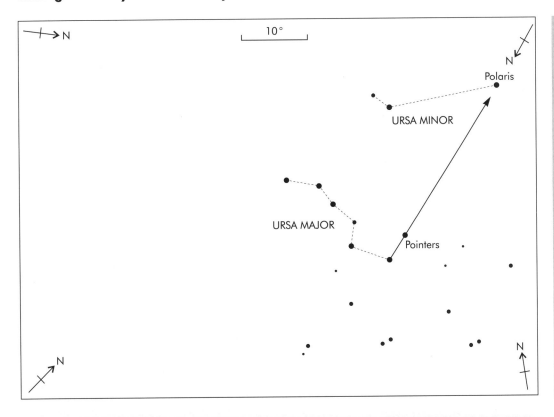

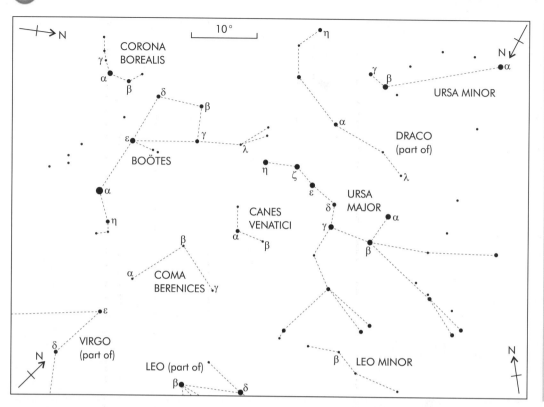

Figure 1.13 All the constellations in the Ursa Major region. These star maps showing all the constellations in the region have the stars marked in four brightness ranges, and a limiting magnitude of +4.5^m, except for a very few faint constellations where fainter stars are needed for the constellation to be shown at all.

Figure 1.14 The Orion region – the stars visible from a typical urban, light-polluted site.

Figure 1.15 The Orion region – the stars visible from a good site.

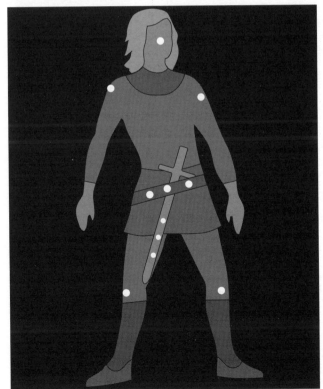

Figure 1.16 Orion as the Hunter.

Figure 1.17 The Orion region – stars visible from a brilliant site to an acute observer.

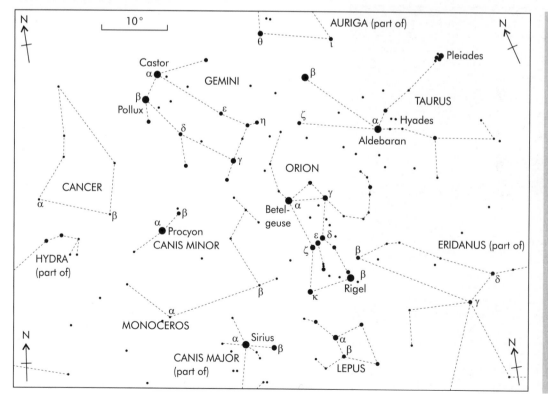

Figure 1.18 All the constellations in the Orion region.

Figure 1.19 The Centaurus region – the stars visible from a typical urban, light-polluted site.

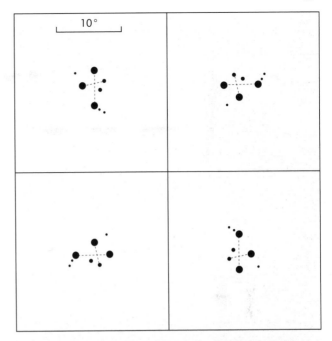

Figure 1.20 The changing appearance of Crux at different orientations.

the telescope more accurately at the estimated position. The finder telescope will often have cross-wires visible through the eyepiece, to enable the pointing to be accomplished as accurately as possible. Once you have seen the object in the finder, you move the whole telescope until the object is centred on the cross-wires, and it should then be visible in the main telescope. Be aware, however, that the image seen through an astronomical telescope is reversed, and therefore the direction moved by the telescope will be opposite to the direction of motion seen through the eyepiece. This is confusing at first, but a little experimentation will soon make you familiar with the phenomenon.

1.4 Positions in the Sky

As we have seen in the previous sections, it is quite possible to find your way around the sky and to acquire faint objects in your telescope without knowing anything about the absolute positions or coordinates of objects in the sky. Indeed, if your telescope does not have any

field of view (Sections 2.2.1 and 2.2.2). A little scanning around the area will then often locate the object quickly. This process is aided if you have a finder telescope attached to your main telescope (see Fig. 2.21 and Section 2.8.4). This is a smaller telescope with a low magnification and a wide field of view, which is aligned with your main telescope. You use it to find the object initially, or to point

Figure 1.21 The Centaurus region – the stars visible from a good site.

Figure 1.22 The Centaurus region – the stars visible from a brilliant site to an acute observer.

means of determining the position that it is pointing to in the sky, then you have no choice but to star hop your way round. Even if your telescope does have scales (or setting circles) on its axes, you do not have to use them, and you can skip over this rather technical section if you so wish.

However, if your telescope has setting circles (see Fig. 1.25, *overleaf*), or is one of the modern computer-

controlled instruments, then you may wish to understand how these work. This will also have the advantage that you can look up the positions of objects in catalogues or on star maps, and set directly on to them, without having to work out a route whereby you may star hop from a known object to the one you want.

The position of objects in the sky is given by a pair of coordinates that are similar to latitude and longitude on

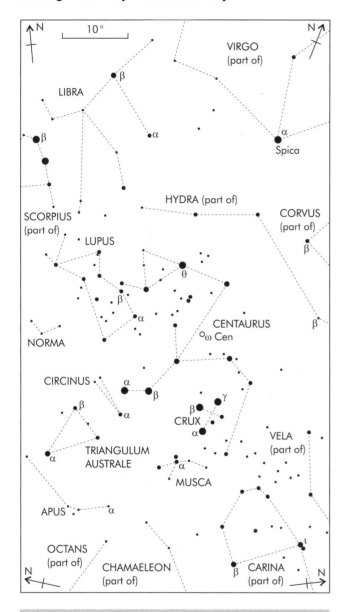

Figure 1.23

Figure 1.23   All the constellations in the Centaurus region.

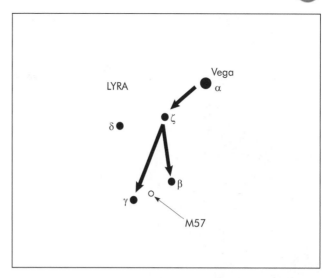

Figure 1.24   Star hopping to the Ring Nebula (M57) in Lyra. The bright star Vega (α Lyr) is easy to find. A short (2°) straight-line hop takes you to ζ Lyr and then two slightly longer hops to γ and β Lyr. M57 is then about half-way between γ and β Lyr.

the Earth (see Fig. 1.26, *overleaf*). The equivalent of latitude is called *Declination* and the equivalent of longitude is called *Right Ascension* (usually abbreviated to Dec and RA, or to their commonly used symbols: α for RA and δ for Dec). Declination therefore measures the angle of an object above or below the equator. Right ascension measures the angle around the equator (see Fig. 1.27, *overleaf*).

There are, however, three complications with RA compared with longitude, which often cause difficulties. The first is where do we measure it from? Longitude is measured from the site of the old Greenwich observatory in London, and there is a fixed strip of metal in the ground to provide the zero point. We therefore need a point in the sky which is the equivalent of this strip of metal to provide a zero for RA. The point that is chosen is where the Sun in its yearly journey around the sky passes from the southern hemisphere to the northern. It is called the First Point of Aries or the *Vernal Equinox*, though it has now moved away from the constellation of Aries into Pisces. It clearly lies on the equator, and the Sun passes through it around 21 March each year. Fortunately the RA setting circle can be set up without knowing exactly where the First Point of Aries is in the sky. The second problem is that although RA is an angular measure like longitude, the units it is measured in are Hours, Minutes and Seconds of time. The reason for this is allied to the third problem, which is that because of the Earth's daily rotation, the heavens appear to rotate around us. Thus if a fixed telescope is directed towards a point in the sky with an RA of say, 6 h 30 m, at one moment, then five minutes later it will be pointing to an RA of 6 h 35 m, and so on. This in fact is just another way of saying that if you look through your telescope with the drive turned off, then the object will soon drift out of the field of view. The Earth rotates

16

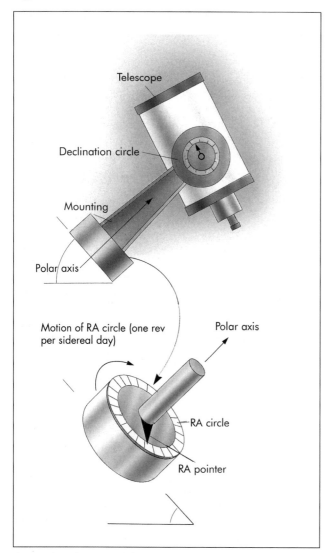

Figure 1.25 Setting circles.

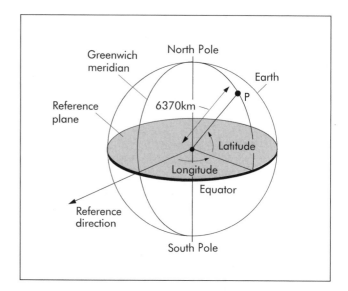

Figure 1.26 Latitude and Longitude.

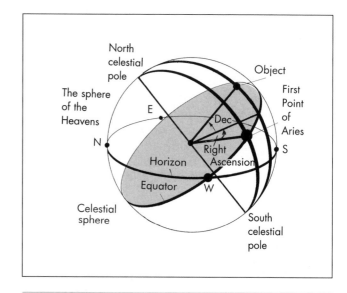

Figure 1.27 Declination and Right Ascension.

once a day, and so 24 h corresponds to 360° in normal angular measure. The interrelationship is thus given by

1 h = 15°	1° = 4 m
1 m = 15′	1′ = 4 s
1 s = 15″	1″ = 0.0667 s

(1.1)

An additional complication is that because of the Earth's orbital motion around the Sun, 24 hours of civil time (which is based on the average time taken for the Sun to return to the same place in the sky) is not the rotational period of the Earth. In 24 hours of civil time, the Earth actually rotates through an angle of 360.986°. The true rotational period of the Earth is thus 23 h 56 m 4 s. This latter period is called the *Sidereal Day*, and is the day based on the time it takes a star, or other fixed object, to return to the same place in the sky. There is also a time system called *Sidereal Time* which is based on the sidereal day. Details of sidereal time, however, are beyond the scope of this book, but the interested reader is referred to other sources listed in Appendix 2. For many purposes, especially to begin with, solar time, as given by normal clocks and watches, may be used without any problems.

Using RA and Dec to set your telescope on to an object is a lot simpler than the above description may have made it appear. This is because the complication of using hours, minutes and seconds for RA is not really a problem because both the RA setting circle (see Fig. 1.25) and the star catalogue or star map will be calibrated in these units, so you do not have to convert to degrees, but simply use the figures directly. Secondly so long as you know the position of any object that you can find directly, you do not need to know where the First Point of Aries is in the sky. Provided that the mounting of your telescope is reasonably well aligned (Section 2.4.1), then the steps to finding an object from its position are therefore as follows:

1. Look up the position of the object in a star catalogue, planetary ephemeris, or read it off a star map. Also find the position of another bright object in the sky that you can find directly.

2. Find that bright object and centre it in your telescope.

3. Move the RA setting circle until it reads the value for the bright object (the Dec value should not need adjustment unless it is mis-aligned).

4. Move the telescope, without moving the RA setting circle, until the RA is correct for the object you are trying to find.

5. Move the telescope until the Dec is correct for the object that you are trying to find.

6. Look through your telescope or its finder, and if you have set things correctly, the object you are trying to find should be visible.

With a computer-controlled telescope, there will be detailed instructions from the supplier on how to set it up. Normally, though, it will still require you to find a bright star at the beginning of the observing session in order to set up the position of the telescope. Thereafter, depending on the system in use, you may just type in the position of the object or even just its name, and the computer will drive the telescope to the correct position.

The RA setting circle (see Fig. 1.25) moves with time in order to retain its alignment as the sky moves (actually it is the Earth that is moving, but it is simpler at times to think of the sky rotating around the Earth). With some telescopes, there may be a setting circle that looks like the RA circle, but which is not driven. This is called an *Hour Angle* setting circle (abbreviated HA) – it simply gives the position of the telescope towards the east or west and cannot be set up so that you can point the telescope directly on to the RA of an object. It is possible to use the HA setting circle to set the telescope, but this involves the use of sidereal time (see above), and is beyond the scope of this book (for further information, see Appendix 2). If you have such a telescope, however, the Declination setting circle may still be used. A combination of star hopping and setting the telescope will probably be quicker than star hopping by itself. With such an instrument, you should move the telescope until the Declination setting circle reads the correct value for the object you are trying to find. Then, without changing the Declination, rotate the telescope around the polar axis until it is pointing roughly in the right direction, as judged against the constellations. Next, while looking through the telescope or its finder, sweep the telescope back and forth around the polar axis, again without changing the Declination, and you should quickly find the object you are looking for.

A final complication needs to be mentioned before we finish with this section. The position of the First Point of Aries is not fixed on the sky, but moves slowly. It takes about 26 000 years to travel right the way around the equator. It is as though the metal strip in the ground at Greenwich from which we measure longitude were to be mounted on the back of a rather slow tortoise which crawled towards the west by 2.5 metres per day! This movement of the First Point of Aries is called *precession*, and it changes the values of RA and Dec of all objects in the sky by up to 50″ per year. But precession does not change the relative positions of all objects. Provided, therefore, that you take the position of your initial bright alignment star and that of the object you are trying to find, from the *same* source, and follow the setting procedure given above, precession can be ignored. You will find, however, that most positions of objects will be given a date. This date is called the *epoch* of the star catalogue or star map, and is the time when the catalogue or map gives correct values for RA and Dec. Epochs are usually chosen at twenty-five years intervals, with 1950, 1975 and 2000 the ones mostly in use at the moment. RA and Dec are often written with the epoch as a subscript, as in RA_{2000} etc., to show this.

1.5 Star Charts and Other Helpful Items

If you have read this far, then you are probably now anxious to go out and use your telescope to tour the exotic further reaches of the Universe. How do you find out where the best and most interesting objects are? The first step is simply to carry on reading this book. Chapters 3 to 10 show many objects within the reach of even the smallest telescope. More importantly, those objects are shown as you should be able to see them. This is an important point, because many people become interested in astronomy through reading books. Often the images in those books are obtained using exposures of many hours on very large telescopes. A small telescope used visually is going to show them very differently.

Once you have found some or all of the objects listed in this book, there are many other sources of information. There are maps of the sky or star charts, some of which are large posters, others in the form of books, yet others are programs which can run on a small computer. These latter are often able to display the sky as it is seen from a particular point on Earth for a particular time of the night and for a particular date. Some may also be able to plot the positions of the Moon and planets. Most of these forms of sky map will indicate interesting objects by special symbols and/or by a list. They are widely available through book shops and astronomy societies, and are advertised in the popular astronomy magazines (see Appendices 1 and 2). The companion book to *Seeing Stars*, *Photo-guide to the Constellations* by Chris Kitchin,

Springer-Verlag, 1997, also shows the positions of many of the most fascinating objects to be found in the sky.

Catalogues of objects are also available. These will usually list objects by type and give their positions and other data on them. Catalogues often cover many more objects than may be seen on a star chart. The *Hubble Guide Star Catalogue*, for example, contains nearly twenty million objects. Some catalogues are available from the same sources as star charts, but they can be quite expensive. The more specialised catalogues can be difficult to find, and are sometimes published as scientific papers in the research journals. If you get to the stage of needing this type of catalogue, then you will probably need to contact your national astronomy society, a good library or a professional observatory. Some of the computer-controlled telescopes have internal databases, listing the positions of thousands of objects, and may also be able to predict the positions of the planets.

The planets, asteroids, comets and ephemeral objects like novae and supernovae pose a problem because of their changing positions and/or unexpected appearance. The positions of the planets are widely available. They are given in the popular astronomy magazines, in some newspapers which publish monthly astronomy columns, in books such as the *Yearbook of Astronomy*, in almanacs produced by national societies and observatories, and via computer programs which are able to predict their positions from the properties of their orbits. Some asteroids and short-period comets will have their positions given in the same sources. For new comets, novae, supernovae etc., you will need to look in the popular astronomy magazines, the newspapers, or join a society which publishes a newsletter. If you have access to the World Wide Web or the Internet, then the astronomy news pages will give details of new objects such as these (Appendix 7).

Chapter 2
Your Telescope and How to get the Best Out of It

2.1 Telescope Designs

What is the best design of telescope? Ask that question of a dozen astronomers and you will get twenty answers – all of them correct! That is because the answer depends on so many factors: what type of objects you wish to observe, whether or not you wish to obtain photographs or charge-coupled device (CCD) images, what your observing site is like, whether you use the telescope on every clear night or just occasionally, whether the telescope is permanently mounted in a dome or has to be brought out of storage when you want to observe, and by no means least, on how much you can afford to spend. Without this last constraint, all amateur astronomers would probably have a Hubble space telescope or a Keck telescope of their own.

This book is intended primarily for people who already have a telescope, so in some ways the above question is already answered; the best design of telescope is the one you have got. However, if you are still at the stage of deciding on a design, or are looking to replace your existing telescope, then it is worth spending some time thinking about what you want before making any commitments. The properties of the main types of designs of telescopes are outlined below, then in Table 2.1 (later in the chapter) their pros and cons for various purposes are summarised. Generally, the larger the aperture, the better, and so cost is usually the deciding factor in the choice of a telescope. But aperture is not the only consideration: a smaller telescope on a good mounting with a drive to enable it to track objects in the sky may be far more use than a huge instrument that has to be moved by hand every few seconds. A good drive is essential if you intend to take photographs or CCD images. Large apertures and small focal ratios are good for observing nebulae and galaxies, but a refractor with a long focal ratio may well be better for planetary observations. Before spending money on an unsuitable telescope, try to sample a number of different designs and makes. A good way of doing this is to join your local astronomical society (Appendix 1), where the enthusiasts will be only too delighted to show you their prized instruments. Finally, do not overlook the possibility of purchasing a second-hand telescope. These will usually be about 50 per cent of the cost of a new instrument, and provided that they have been cared for, will be just as good. The popular astronomy magazines (Appendix 2) usually have small advertisement sections where telescopes are offered for sale.

2.1.1 The Refractor

The first telescope known to have been pointed at the sky (by Galileo in 1609) was a simple refractor with a converging lens at the front and a diverging lens near the eye. It gave upright images at a low magnification (×30 was the best that Galileo achieved). A *refractor* is the name for any telescope in which the main light collection is due to a lens, rather than to a mirror (the latter produces a *reflector* – see below). Galileo's telescope and other designs based on simple lenses suffered from a major drawback. It was impossible to focus the image properly because light of one colour was brought to a focus at a different point from that for light of a different colour.

Thus the blue image of an object would be in focus at one point but would be surrounded by a haze produced by the out-of-focus green, yellow and red images, and so on. This problem is known as *chromatic aberration* (see Fig. 2.1).

Chromatic aberration still occurs in today's refractors, but it is much reduced and most of the time is no longer a problem. John Dollond in 1754 discovered that he could make a compound lens from two simple lenses. If he chose the right varieties of glass for the lenses, then the chromatic aberration of one lens almost cancelled out that from the other. The compound lens is called an *achromatic doublet* (see Fig. 2.2). For large refractors, such as the 1 m Yerkes telescope, the residual chromatic aberration with an achromat is still a problem; but for telescopes with focal lengths of one or two metres, modern achromats are very effective. It is possible to add more lenses to improve the colour correction further, and this is usually done for cameras. But for the large-diameter lenses used in telescopes, adding more lenses to the basic achromat usually becomes far too expensive.

Refractors are usually more expensive than reflectors (see below) of similar diameters. This is because with the achromatic doublet there are four surfaces to be ground and polished instead of the one required for a mirror. However, refractors have advantages for some purposes which may make it worthwhile spending the extra

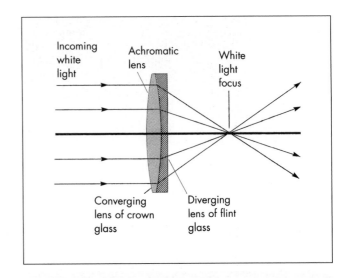

Figure 2.2   The basic achromatic doublet.

money. The main advantage is that a refractor does not have the central obstruction due to the secondary mirror which reflecting telescopes require. This means that, other things being equal, refractors can produce better images than reflectors. In particular, their images lack the spikes often to be seen in images produced by reflectors (see Fig. 2.3), and which arise through diffraction around the supports for the secondary mirror. When looking at extended images, such as the Moon and planets, the effect of these spikes is to reduce the contrast in the image produced by a reflector compared with that produced by a refractor. Thus if planetary observation is to be your main area of interest, then a good-quality refractor is the telescope to choose. Additional minor advantages of refractors are that their optics are usually more firmly mounted than in reflectors and so remain collimated (see Section 2.3) better, and the sealed tube keeps the surfaces of the lenses cleaner and tends to reduce convection currents.

A word of warning! Many places, including mail-order firms, chain stores, camera shops etc. sell small refractors at comparatively low prices (see Table 2.1, later in the chapter). These typically have an apparent aperture of a few centimetres. They are often mounted on a ball-and-socket pivot on a tripod (see Fig. 2.4, *overleaf*). Such telescopes are usually not worth considering. They will often have stops inside the telescope tube (visible by looking through the main lens) because the optical quality of the lenses is poor, and these reduce the effective aperture

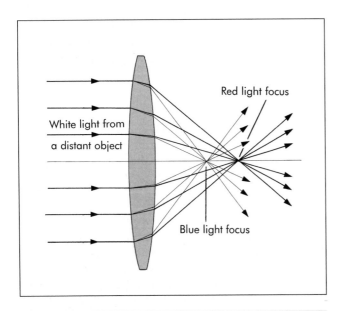

Figure 2.1   Chromatic aberration in a simple lens.

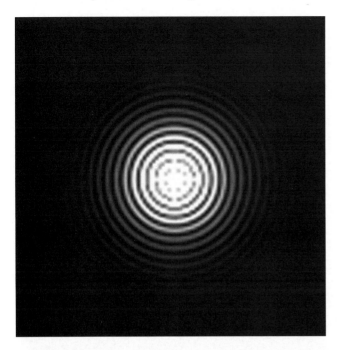 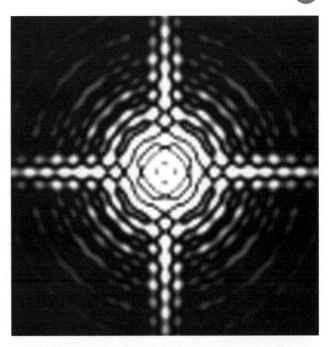

Figure 2.3 Images of stars produced by a refractor (*left*) and a reflector (*right*), showing the diffraction spikes produced by the secondary mirror supports of the reflector.

Table 2.1. How to choose a telescope

Telescope[a]	Cost[b]	Portable?[c]	Observing programme(s)[d]
Binoculars 7 or 10 × 50 hand-held	£75–£300 $100–$500	Yes	*Visual* – Moon and planets (but will not show much detail), comets, brighter nebulae and galaxies
Binoculars 20 to 25 × 75 to 100, on a mounting	£200–£500 $300–$700	Yes	*Visual* – Moon and planets – will start to show some details, comets, nebulae and galaxies
Home-made 15 cm, f8 Newtonian Reflector on a Dobsonian mount	£0–£200 $0–$300	Yes	*Visual* – Moon, planets, double stars, brighter nebulae and galaxies
4–6 cm Refractor, or a 7–12 cm Newtonian Reflector, on a pivot-type mounting or unmounted, probably purchased from a mail-order company or a chain store	£25–£100 $40–$150	Yes	Bottom of the dustbin (which is where you will throw it soon after buying it!)
7.5 cm, f14 Achromatic Refractor on an undriven equatorial mounting	£300–£500 $400–$700	Yes	*Visual* – Moon, planets, double stars, nebulae and galaxies only with difficulty
15–20 cm, f8 Newtonian Reflector on a driven equatorial mounting	£400–£1000 $600–$1500	Just about	*Visual* – Moon, planets, double stars, brighter galaxies and nebulae *Imaging* – photography of all of the above, CCD images of fainter nebulae and galaxies as well *(continued overleaf)*

Table 2.1 (*Continued*)

Telescope[a]	Cost[b]	Portable?[c]	Observing programme(s)[d]
15 cm, f10–f12 Achromatic Refractor on a driven equatorial mounting	£1000–£2000 $1500–$3000	Just about	*Visual* – very good for Moon, planets, double stars, micrometry *Imaging* – photography of all of the above, CCD images of brighter nebulae and galaxies as well
40–60 cm, f4–f6 Newtonian Reflector on an undriven Dobsonian mounting	£2000–£5000 $3000–$7000	The adverts say they are!	*Visual* – Moon, planets, double stars, very good for nebulae and galaxies
10–12.5 cm, f10 Schmidt–Cassegrain telescope on an equatorial mounting with a drive	£1000–£2000 $1000–$2500	Yes	*Visual* – Moon, planets, double stars, brighter nebulae and galaxies *Imaging* – a bit lightweight to take a camera or CCD, but possible for Moon, and planets
20–27.5 cm, f10 Schmidt–Cassegrain telescope on an equatorial mounting with a drive	£2000–£3000 $2000–$4000	Yes	*Visual* – Moon, planets, double stars, nebulae and galaxies *Imaging* – photography and CCD imaging of all of the above
20–27.5 cm, f10 Schmidt–Cassegrain telescope, computer-controlled drives and on an alt–az mounting	£2500–£4000 $2500–$5000	Yes	*Visual* – Moon, planets, double stars, nebulae and galaxies *Imaging* – only short exposures with the object near the meridian possible because the field of view rotates with time when using an alt–az mounting
40 cm, f10 Schmidt–Cassegrain telescope, computer-controlled drives and on an alt–az mounting with an image de-rotator, or on a driven equatorial mounting	£15 000–£17 000 $15 000–$17 000	No	*Visual* – Moon, planets, double stars, nebulae and galaxies *Imaging* – complete range of observations possible up to research standard

[a] These are typical examples of instruments that may be found on offer. Slight differences in particular cases (such as f12 instead of f14, or 18 cm diameter instead of 15 cm) will make little difference to the suitability of the instrument.

[b] The cost here is the UK/USA prices for new instruments at the time of writing (1996). Perfectly good second-hand versions of these telescopes may be found at about 50 per cent of these values.

[c] This depends on how strong (and determined) you may be, and also on whether the instrument can be broken down into sections.

[d] This will also depend on the eyepieces that you have available, and the quality of your observing site. For imaging, it will also depend on how good the tracking of the telescope is, and for how long you are prepared to guide an exposure. The Sun has been deliberately omitted from all the programmes, and should only be observed if you have suitable filters (see Chapter 3).

considerably. They have very small fields of view, making it very difficult to find and retain an object in the eyepiece, and finally the stands are too flimsy to be of any use at all. More budding astronomers have probably been put off the subject by being presented with such an instrument by a doting aunt or uncle and then finding it impossible to use, rather than by all the freezing nights and clouds coming up at the wrong moment put together.

2.1.2 The Newtonian Reflector

The telescope of choice for many people is the Newtonian Reflector, because it is usually the cheapest way to purchase a telescope with a reasonable aperture. As the name suggests the primary optical component is a mirror, and the design originated with Isaac Newton in 1668. The mirror is concave, and has a reflective coating, usually of aluminium, on its front surface so that the light does not

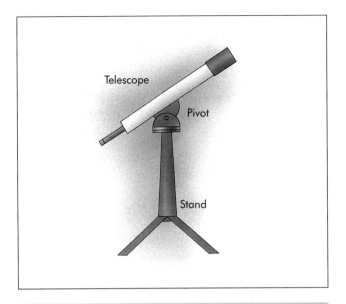

Figure 2.4 A telescope to avoid!

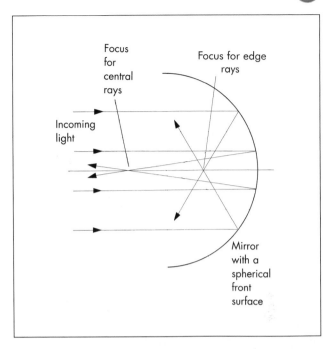

Figure 2.5 Spherical aberration in a mirror.

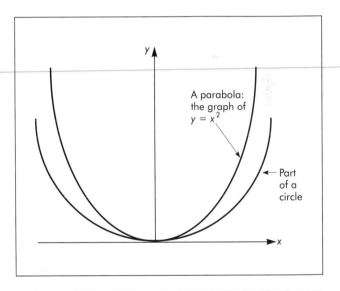

Figure 2.6 A parabola: a deeper curve than a circle.

have to pass through any glass. The use of a mirror in place of a lens immediately eliminates the problem of chromatic aberration because light of one colour is reflected in the same manner as light of any other colour. However, if the surface is spherical, then another problem arises – that of *spherical aberration* (see Fig. 2.5). In this fault, the light near the edge of the mirror is brought to a different focus from that falling near the centre of the mirror. The shape of the surface then has to be deepened to that of a parabola (see Figs 2.6 and 2.7, *overleaf*), in order to eliminate spherical aberration. Finally a small flat mirror is required to reflect the light to the side, so that your head does not block all the light going into the instrument (see Fig. 2.8, *overleaf*). The secondary mirror does block out a small amount of light. but this is less important than the deterioration introduced into the image by diffraction around the mirror and its supports (see Fig. 2.3).

The Newtonian telescope is usually a relatively cheap instrument because of the simplicity of its design. It does however have some problems. In order to keep its size down, the focal ratio of the primary mirror is often quite small: f8 or f6, or even less. This means that the image quality will deteriorate quickly as you go away from the centre of the field of view. For visual work this is not too important, because you can always bring the bit of the object that you are looking at to the centre of the eyepiece. However, if you use the telescope for photography or for CCD[1] imaging, then the outer parts of the image will often appear fuzzy and out-of-focus. The eyepiece is high on the

---

[1] At present, the small physical size of the CCD chips generally used outside major professional observatories makes this problem not too important. However, as CCD detectors become larger, it will become of increasing significance.

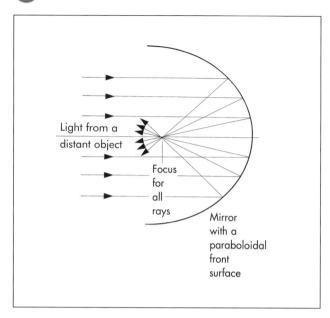

Figure 2.7 A parabolic mirror.

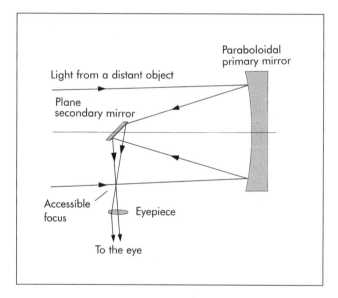

Figure 2.8 The Newtonian telescope.

The simplicity of the Newtonian telescope means that it is quite possible for those with moderate practical skills to make such telescopes for themselves. This is then by far the cheapest way to acquire a telescope. Depending on how much of the construction you undertake (and on how good you are at scrounging!), the cost of a respectably sized telescope may be reduced to close to zero. Books describing in detail how to make a Newtonian telescope, including how to grind and polish the mirrors, are listed in Appendix 2.

2.1.3 The Cassegrain Reflector

The difference between the Newtonian and Cassegrain designs of telescope lies in the secondary mirror. In the Cassegrain telescope this has a convex hyperboloidal surface (see Fig. 2.9), and the light is reflected back towards the primary mirror and out through a hole in the centre of that mirror (see Fig. 2.10). The Cassegrain design, and its derivative the Ritchey–Chretien, is almost universally used for large telescopes. However, for smaller instruments its advantages do not normally outweigh its greater cost compared with the Newtonian design. It is therefore very rare to come across a Cassegrain telescope of smaller size than about 0.4 or 0.5 m. The advantages of the Cassegrain design are that the eyepiece is at the back of the telescope, where it is more easily accessible, and that the secondary mirror expands the focused beam from the primary mirror to give a much longer focal ratio. The

side of the telescope, which can make it difficult to balance properly, and with the larger instruments you will need a step ladder to reach it. Finally, the mirrors will frequently need to be realigned (collimated – Section 2.3)

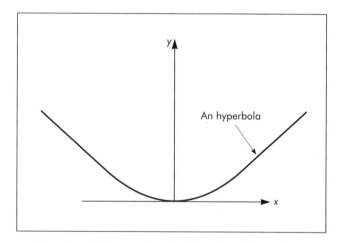

Figure 2.9 The hyperbola.

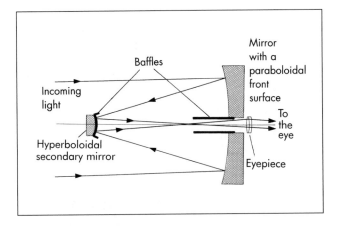

Figure 2.10 The Cassegrain telescope.

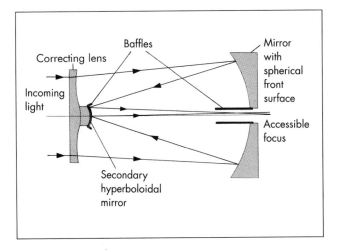

Figure 2.11 The Schmidt–Cassegrain telescope (the name derives from the use of the correcting lens and spherical primary mirror in a similar way to the correcting lens and mirror in a Schmidt camera).

effective focal length of a Cassegrain telescope is thus much greater than its physical size. For a given focal length, the Cassegrain telescope is much shorter than the Newtonian, making its mounting and dome etc. smaller and cheaper. The Cassegrain telescope and its relatives need carefully designed shields (called baffles) however, to prevent light reaching the eyepiece directly from the sky.

2.1.4 The Schmidt–Cassegrain Telescope

Although small Cassegrain telescopes are rare, a variation known as the Schmidt–Cassegrain Telescope (SCT) has become very popular in the last couple of decades. There are several firms making SCTs with apertures ranging from 12.5 cm to 40 cm, and because of the large number of such instruments being produced, their prices are remarkably good value, though still considerably higher than that for a Newtonian telescope of the same aperture. The design of the telescope is similar to that of the Cassegrain (see Fig. 2.10), but with the addition of a weak lens (see Fig. 2.11) and the use of a spherical primary mirror. The secondary mirror is attached to the lens, eliminating the support arms and their diffraction spikes (see Fig. 2.3). The lens has a complex shape designed to eliminate many of the faults (aberrations) of the basic Cassegrain design and of the spherical mirror, but is too thin to introduce any significant chromatic aberration of its own. The primary mirror can therefore be made with a

very short focal ratio, and the telescope is very compact for its effective focal length. A typical 0.2 m f10 SCT, with an effective focal length of 2 m, would be physically only some 40 cm long. The telescopes are therefore light, easily mounted and very portable. They also have excellent image quality. A number of firms now supply them with computer-controlled drives, enabling them to be used on either equatorial or alt–azimuth mountings. The computers have the positions of many thousands of the more interesting objects in the sky stored within them, and can drive the telescope to centre on a selected object within a few seconds of time. Such telescopes eliminate much of the frustrating time and effort that will be required with other designs of telescope to find objects in the sky, especially the fainter and more elusive ones. But they also reduce the skill and the sense of achievement that you get when you do find a difficult object for yourself. Nonetheless, a computer-controlled SCT is probably the best all-round small telescope currently available and would be most people's first choice, provided only that they can afford one.

2.2 Eyepieces

The objective lens or mirror of the telescope produces a focused image which may be seen on a sheet of cardboard,

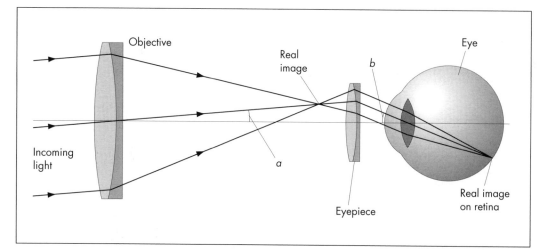

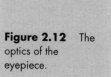

Figure 2.12 The optics of the eyepiece.

or used to produce a direct photograph or CCD image. Contrary to many people's expectations, the eyepiece does not produce an image. Instead it converts the conical beam of radiation from the objective into a parallel beam, and the eye then focuses this beam on to the retina to produce the image that we see when looking through the telescope (see Fig. 2.12). By moving the eyepiece from its normal position it is possible to produce a real image, and this is used for solar (Chapter 3) and planetary work (Chapter 5) when it is known as eyepiece projection.

2.2.1 Magnification

It is important to understand the correct role of the eyepiece, because otherwise the magnification of the telescope does not make sense. Clearly, when looking at an object such as the Moon, which is 3476 km across, through a telescope with a magnification of, say, ×80, we do not have a resultant image which is 278 080 km across. Magnification for a visually used telescope is thus the increase in the angular size of the object, not the increase in its linear size, and it is given by the ratio of the angles b to a in Fig. 2.12. A simpler to use formula for the magnification however is

$$\text{Magnification} = \frac{\text{Focal length of the objective}}{\text{Focal length of the eyepiece}} \quad (2.1)$$

Thus a 20 cm, f10 telescope (focal length 2 m) used with a 25 mm focal length eyepiece would give a magnification of ×80, and so on.[2]

Magnification is widely regarded as the most important function of a telescope. An eyepiece with a short focal length may thus be treasured over longer focal length eyepieces. However, high magnification is often not needed and may give poorer results than lower magnifications. This is because the Earth's atmosphere limits the quality of what we can see through a telescope via the effects of twinkling (also called the "seeing" and "scintillation"). Save under exceptional observing conditions, magnifications of ×150 or ×200 will be as much as can be used. Higher magnifications will give a larger image but one in which less detail can be seen because the contrast will be reduced. Only for double star work will magnifications much higher than these figures be useful. If a telescope is to reach its diffraction-limited resolution (Section 2.5), then a magnification of around ×1300D is needed (where D is the diameter of the objective in metres). Thus a 20 cm telescope would need a magnification of ×260 to enable its theoretical resolution of 0.6″ to be reached.

2.2.2 Other Properties of Eyepieces

Magnification is only one of the important parameters of an eyepiece. As with the objectives, the eyepiece must be

---

[2] Eyepieces from old microscopes can frequently be rescued and used with your telescope. Such eyepieces will usually have a magnification such as ×10 marked on them. This bears no relationship to the magnification that the eyepiece will give when used on the telescope. That must be calculated using Eq. (2.1). The focal lengths of microscope eyepieces are: ×10 – 25 mm, ×15 – 17 mm, ×20 – 12.5 mm, and ×25 – 10 mm.

of adequate optical quality, or images will be degraded. Thus a simple converging lens could act as an eyepiece, but it would introduce chromatic aberration (Section 2.1) into the image. Most eyepieces are therefore compound devices, containing two or more lenses. Individual designs are discussed in Section 2.2.3.

A simple lens can accept light from a very wide range of angles. The compound designs used for actual eyepieces, however, are much more restricted. The size of the cone of emerging rays from an eyepiece is called its *field of view* (see Fig. 2.13). The field of view of the telescope is then the field of view of the eyepiece divided by the magnification. Thus an eyepiece with a field of view of 45°, used at a magnification of ×200, would enable a circle of the sky 13.5′ across to be seen. For a given type of eyepiece, the higher the magnification, the smaller the field of view of the telescope. For angularly large objects such as star clusters, nebulae, the Moon etc., a low magnification will thus be needed in order to give a large enough field of view to see the whole object. Similarly when finding objects, the lowest power eyepiece is generally the best, since it gives the widest field of view, and the best chance of locating whatever is being sought.

An important criterion for being able to use an eyepiece comfortably is the *eye relief*. This is the distance from the final lens of the eyepiece to the eye. In order to get the best view through an eyepiece, the pupil of the eye needs to coincide with an area known as the exit pupil, through which all the light from the eyepiece passes (see Fig. 2.13).

Ideally the eye relief should be about 6–10 mm. If it is less than 6 mm, then the eyeball will be uncomfortably close to the lens (there have even been cases of astronomers getting frozen to the eyepiece on cold nights!). More than 10 mm, and it will be difficult to keep your head steady enough to see the whole field of view.

2.2.3 Choosing an Eyepiece

Eyepieces come in three types: Expensive, Very Expensive and Ouch! (Classes A, B and C in Fig. 2.14, *overleaf*). Since you will need several eyepieces for your telescope to cater for different observing conditions and different objects, it is important to choose wisely or you will be wasting considerable amounts of money. The main difference between the types of eyepieces is in their fields of view. An eyepiece at the least expensive end of the scale would have a field of view of 35° to 45°, a wide angle eyepiece would have a field of view of 55° to 65°, while for an ultra-wide angle eyepiece it might be up to 85°. The prices would correspondingly range (1996 values) from £30 to £100 ($50 to $200), through £60 to £200 ($100 to $300) to £150 to £250 ($200 to $400). There are numerous different designs, some dating back to the early days of telescopes, others computer-designed and using exotic glasses to give optimum results. The main varieties are shown in Fig. 2.14 (*overleaf*), though there is considerable variability in the details of the designs between different manufacturers.

How to choose an eyepiece? First consider how you use your telescope. It is likely that you start off by searching for the object you are interested in. For this you need the widest possible field of view to give yourself the best chance of coming across it. That means a low-power, wide-angle eyepiece. However if the magnification is less than about ×140D, or ×28 for a 20 cm telescope (where D is the diameter of the objective in metres), then not all the light gathered by the telescope will enter the eye, so very low powers should be avoided. Having found the object, you will then want to observe it. For angularly large objects, such as M31 or the Moon, you may well continue using the low power so that the whole object is visible. For angularly smaller objects such as a lunar crater, a planet or a double star etc., you will probably centre the object in the field of view and then exchange the low-power eyepiece for one with a greater magnification; perhaps then centring in this eyepiece and going on to an even higher power. These higher power eyepieces do not need to have wide fields of view, because centring each time in the lower power eyepiece will ensure that the object is visible at the higher

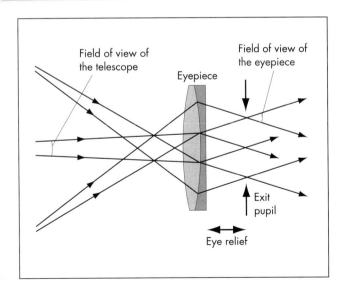

Figure 2.13 The field of view of an eyepiece.

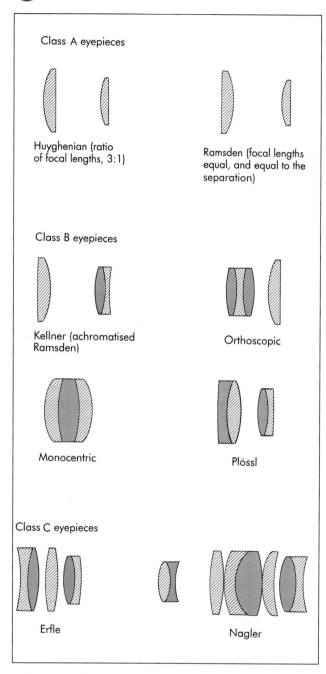

Class A eyepieces

Huyghenian (ratio of focal lengths, 3:1)

Ramsden (focal lengths equal, and equal to the separation)

Class B eyepieces

Kellner (achromatised Ramsden)

Orthoscopic

Monocentric

Plössl

Class C eyepieces

Erfle

Nagler

Figure 2.14 Types of eyepiece designs.

observe it, and most designs of eyepiece give good images close to their optical axes. Thus a good choice of eyepieces would be a low-power, wide-angle eyepiece, and two or three higher power eyepieces of a cheaper design. The very simple Huyghenian and Ramsden (see Fig. 2.14) eyepieces should not be considered, so that for say a 20 cm, f10 telescope, ideal eyepieces could be:

a wide-angle 40 mm eyepiece giving ×50
a Plössl or Orthoscopic eyepiece of 25 mm focal length, giving ×80
a Plössl or Orthoscopic eyepiece of 12.5 mm focal length, giving ×160
and a Plössl or Orthoscopic eyepiece of 8 mm focal length, giving ×250

Finally if you change eyepieces frequently, then it is worth considering a parfocal set of eyepieces. These all have their foci at the same spot, so that when you change an eyepiece there is no need to re-focus the telescope. Such a set of eyepieces may most conveniently be mounted in a carousel so that they may be changed very rapidly just by rotating the carousel head from one eyepiece to the next.

A range of magnifications may be obtained with just one eyepiece if it is used with a Barlow lens. The latter is a negative lens placed in front of the eyepiece. If positioned correctly it has the effect of reducing the converging angle of the cone of rays from the objective (the secondary mirror in a Cassegrain telescope fulfils the same function). This is the equivalent of increasing the focal ratio, and so the focal length of the objective. The overall magnification (Eq. (2.1)) is therefore increased. By changing the distance of the Barlow lens from the eyepiece a range of magnifications can be obtained. Unless the Barlow lens is large, however, the combination will usually have a small field of view, and so it does not necessarily replace the need for a range of eyepieces.

A useful accessory for eyepieces is the star diagonal. This is an adapter which goes between the eyepiece and the telescope and which reflects the light through 90°. It saves much strain on your neck muscles when observing near the zenith. With all except the Newtonian design of telescope, observations near the zenith without a star diagonal would necessitate looking vertically, and unless your telescope allows you to lie down while observing, this would require your head to be placed at a very awkward angle. With some mountings (such as the English, Yoke and Fork designs: Fig. 2.17, *overleaf*), it may be impossible to look directly through the telescope when it is pointing near to the Pole. Then a star diagonal

power. Nor do they need to be of brilliant optical quality over the whole field of view, because you will naturally bring the object to the centre of the field in order to

becomes essential to allow the images to be seen at all. Star diagonals however can degrade images slightly, so that their use should be avoided when undertaking critical observations.

2.3 Collimation

The most perfect primary lens or mirror and the best eye-pieces you can afford will fail to produce usable images if they are not properly aligned. Aligning the optics of a telescope is a process called *collimation*. Some telescopes, such as refractors and Schmidt–Cassegrain telescopes, will rarely need collimating unless they have been roughly handled. Others, such as Newtonians which have their secondary mirrors at an acute angle to the optical axis, will need collimating regularly however careful you may be in looking after them.

The basic principle in collimating a telescope is that all the optical components should be centred on the optical axis, and be perpendicular to it (or at 45° in the case of the Newtonian secondary mirror). Fortunately, bringing this state of affairs about does not require you to know where the optical axis is, and be able to measure the angles of lenses and mirrors with respect to it. Instead, the alignment is carried out by looking directly at the lenses and mirror and their reflections. When the optical components are all aligned correctly, the components, their mountings and their reflections when observed through the eyepiece holder without an eyepiece in place, will all be concentric. Collimating a telescope thus requires you to remove the eyepiece, to look into the telescope, and then to adjust the mirror and lens mountings etc. until concentricity is achieved (see Figs 2.15 and 2.16). Most reflectors will have adjusting screws on their primary and secondary mirror mountings to allow you to adjust the mirror positions. Trial and error will soon show you the right direction in which to adjust these screws. Refractors and Schmidt–Cassegrain telescopes may be checked for collimation by looking through their eyepiece holders in the same way. However, if they are out of adjustment, it is probably best to return them to the manufacturer or to a competent optical company for realignment.

2.4 Mountings

Most telescopes sold for astronomical purposes will be provided with a mounting; that is to say, a means of

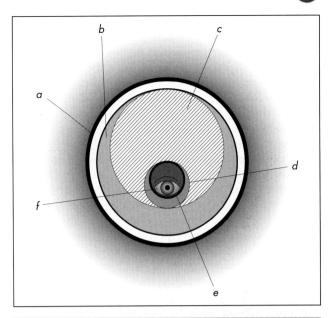

Figure 2.15 The view through the eyepiece holder of a mis-aligned Newtonian telescope: (*a*) eyepiece holder, (*b*) secondary mirror, (*c*) primary mirror reflected in the secondary mirror, (*d*) secondary mirror reflected in the primary mirror and the secondary mirror, (*e*) eyepiece holder reflected in the secondary mirror (twice) and the primary mirror, (*f*) eye reflected in the secondary mirror (twice) and the primary mirror.

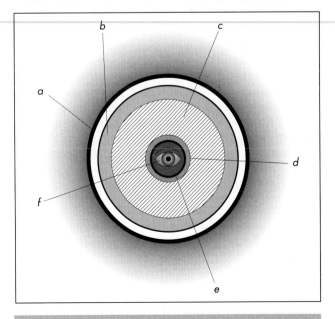

Figure 2.16 The view through the eyepiece holder of a correctly collimated Newtonian telescope. See Fig. 2.15 for key.

holding the telescope steady and pointing towards various parts of the sky. The better mountings will also have slow motions and drives to enable the telescope to track and be guided on an object for a long period of time. At the top end, mountings may incorporate computers which have databases containing the positions of thousands of the more interesting objects in the sky and which may be found at the push of a button.

At the top end of the market (see Table 2.1 earlier in the chapter), mountings are generally excellent and do everything required of them. Some telescopes may now be supplied on alt–azimuth mountings (see below), which are fine for visual work. However over a period of time, the field of view in the telescope when on an alt–azimuth mounting rotates with respect to the telescope. Such mountings are therefore unsuitable if you wish to take long-exposure photographs or CCD images. Field de-rotators may be available from the manufacturer to overcome the problem, but they can be complicated to use. Generally, if you wish to undertake long exposures through your telescope (and most people will do so eventually, if not to begin with), then an equatorial mounting (see below) is to be preferred to an alt–azimuth.

Among the cheaper telescopes (ignoring the very cheap ones, such as that in Fig. 2.4), the optics are often excellent, but savings are made by the manufacturers when producing the mounting. A low-cost way therefore of acquiring a high-quality instrument, if you have a reasonable level of practical skills, is to purchase one of these cheaper telescopes and then to upgrade or replace the mounting yourself. The main requirements in producing a mounting are stability (which usually means substantial components for the mounting), and smooth motions. If it is to be driven, then the drive rate must be correct, and the mounting must be able to be aligned accurately. If you do decide to make your own mounting, then components such as bearings etc. can often be obtained very cheaply from scrap yards as parts of old motor-cycle engines etc.

The two main types of telescope mounting are the *equatorial* and the *alt–azimuth*, and these are considered below.

2.4.1 The Equatorial Mounting

Although there are a number of different versions of this type of mounting, the designs all have one thing in common: one of the axes is parallel to the Earth's rotational axis. This axis is called the *Polar axis*, because if you were to look along it you would be looking at the Celestial Pole. Rotating the telescope around this axis in the opposite direction to the Earth's rotation (tracking) enables objects to be kept in the field of view of the telescope. Only one, constant-velocity motion is required to provide tracking with an equatorial mounting, compared with two variable-speed drives for an alt–azimuth mounting (Section 2.4.2). Equatorial mountings are therefore easily and cheaply provided with drives.

The second axis of an equatorial mounting is perpendicular to the first and is called the *Declination axis*. The name provides a clue to the second reason why equatorial mountings are popular – they are easily related to the positions of objects in the sky (Hour Angle or Right Ascension and Declination – Chapter 1). A movement around the Polar axis changes only the HA or RA of the part of the sky that is being observed, moving around the Declination axis changes only the Declination that is being observed. Position indicators (setting circles) attached to the axes can thus read HA or RA and Dec directly, greatly simplifying the task of finding fainter objects.

Some of the various designs for equatorial mountings are sketched in Fig. 2.17.

A crucial factor in getting an equatorial mounting to work properly is aligning the Polar axis so that it is parallel to the Earth's Rotational axis. Initially this can be done by eye: looking along the axis towards the Pole Star (Polaris). With considerably greater accuracy, if the mounting has setting circles, then the telescope may be set to the coordinates of Polaris, and then by moving the whole mounting and not the telescope itself, Polaris brought to the centre of the field of view. Neither of these methods, however, will give a precise enough alignment to allow long exposures without excessive guiding corrections being needed. Precise alignment therefore requires observations over a period of time and successive adjustments of the mounting. Corrections to the altitude of the Polar axis are found by observing a star between 30° and 60° declination, about 6 hours east or west of the observer's meridian. The star should be centred on the cross-wires of a high-power eyepiece, and tracked using the telescope drive. If, after a while, a star to the east of the meridian is observed to drift northwards[3] in the eye-

---

[3] Remember that astronomical telescopes give inverted images, and that if you are using a star diagonal, the orientation will be changed again. To find the actual orientation (N–E–S–W) in your field of view, first turn off your telescope drive. The stars will drift towards the west. Then gently move the telescope towards the north, and the stars in the field of view will then drift towards the south.

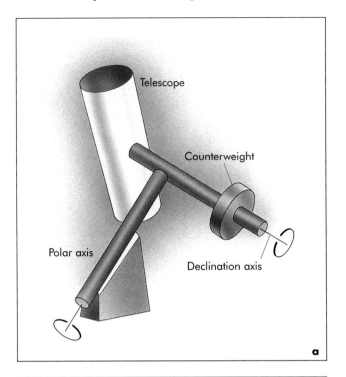

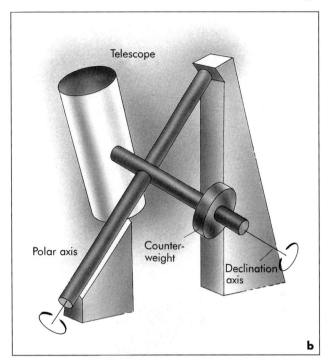

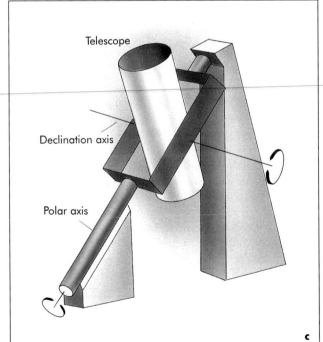

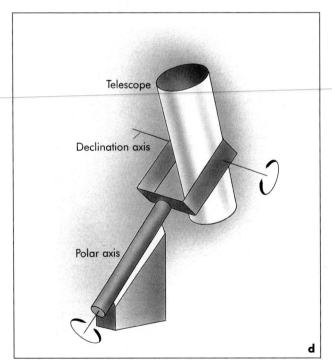

Figure 2.17 Designs for equatorial mountings. **a** German Mounting **b** Modified English Mounting **c** English or Yoke Mounting **d** Fork Mounting

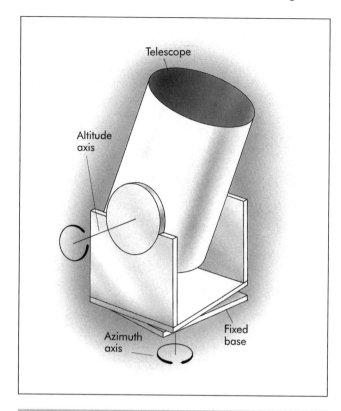

piece, then the altitude of the Polar axis is too high. A star to the west of the meridian would drift to the south in the same circumstances. If the easterly star drifts south, or the westerly star north, then the Polar axis is set to too low an altitude. After a suitable adjustment to the axis, the procedure is repeated, until any remaining drift is acceptable. Alignment will clearly have to be done much more precisely if the observer intends, say, to undertake photography with exposures of several hours rather than visual work. A similar procedure may be used to align the Polar axis in azimuth. Two stars are selected which are ten or twenty degrees north and south of the equator, and differing in right ascension by a few minutes of arc (reference to a reasonable star catalogue will probably be necessary for this – Appendix 2). When the stars are within an hour or so of transiting the observer's meridian, with the drive OFF, the telescope is set just ahead of the leading star (the one with the smaller right ascension). A stop watch is started as that star transits the vertical cross-wire. The telescope is then moved to the declination of the second star, and the time interval until that star transits is measured. The correct time interval between the two transits can be found from the difference in their right ascensions. If the measured time interval is too small, then the Polar axis is aligned to the east, if the time interval is too large, then the Polar axis is to the west of its true direction. Adjustments to the position of the Polar axis are made and the procedure again repeated until satisfactory.

2.4.2 The Alt–Azimuth Mounting

The Alt–Azimuth mounting has one vertical and one horizontal axis. Moving the telescope around one axis changes its altitude, and around the other its azimuth, hence the name for the mounting. Alt–az mountings for small telescopes come in two varieties: very simple and extremely sophisticated. The very simple ones, such as the Dobsonian (see Fig. 2.18) just have the two axes without drives, and the telescope is pushed around by hand. Tracking an object requires the observer to make adjustments to the telescope's position every few seconds or so. It is not possible to take photographs or CCD images with exposures longer than a second or two. It is however possible to place the whole mounting on a drive which, by means of an inclined plane, will enable the telescope to track an object for up to about 10 minutes (see Fig. 2.25). However, by the time you have gone to this amount of

Figure 2.18 The Dobsonian version of the alt–azimuth mounting.

trouble, you might just as well have put the telescope on to an equatorial mounting from the start.

The extremely sophisticated alt–az mountings have only become available for small telescopes relatively recently (they have been used on very large professional instruments for a couple of decades). They have drives in both axes, and the motor speeds are computer-controlled so that the telescope tracks objects correctly. Some versions have databases containing the positions of many thousands of objects. Objects may be selected from the database and the telescope driven automatically to the correct position easily. Even these sophisticated alt–az drives suffer from the drawback that the field of view rotates as the telescope tracks across the sky however. Long-exposure imaging is therefore not possible unless a device is added (an image de-rotator) to counteract the field rotation. Field de-rotators add considerably to the cost of the mounting and are complicated to use. An equatorial mounting is therefore still to be preferred if you wish to obtain long-exposure images.

2.5 Optics

We have already looked at some aspects of the optics of telescopes when considering the various designs of telescope (Section 2.1) and eyepiece (Section 2.2). Here some further aspects of optics that you need to consider when choosing a telescope or optimising an existing telescope are examined.

2.5.1 Light Grasp

The bigger the better? – generally the answer would be yes when choosing a telescope. A major function of any telescope is to gather more light than the unaided eye is able to do. This is called the *light grasp* of the telescope and is simply the ratio of the collecting area of the telescope to the area of the dark-adapted pupil of the human eye. It is given approximately by

$$\text{Light grasp} = 20\,000\,D^2 \qquad (2.2)$$

where D is the diameter of the telescope objective in metres. So the larger the diameter of the telescope, the greater its light grasp; a 75 mm (3-inch) telescope would have a light grasp of 110, a 200 mm (8-inch) telescope a value of 800. The Keck telescopes with diameters of 10 m would have light grasps of 2 000 000 if they were ever to be used visually.

A telescope will increase the brightnesses of point sources, such as stars, by an amount equal to the light grasp. From a good observing site, on a clear moonless night, an observer with reasonable eyesight can see stars of magnitude 6 with the naked eye.[4] The telescope improves this limiting magnitude to

$$m_{\text{lim}} = 16 + 5\log_{10} D \qquad (2.3)$$

---

[4] Astronomers measure the brightness of objects in the sky in terms of magnitudes. The magnitude scale is a geometrical one, so that an object that is one magnitude brighter than another is 2.512 *times* brighter than that second object. The scale is also reversed so that the brighter objects have magnitudes which are numerically smaller than those of fainter objects. The Sun thus has a magnitude of –27, the full Moon –13, Jupiter –2.5, Sirius –1.5, Betelgeuse –0.7, and Polaris +2. For further details see Chapter 7 and books such as *Telescopes and Techniques* by C. Kitchin (published by Springer, 1995).

Thus a 75 mm (3-inch) telescope would have a limiting visual magnitude of 10, a 200 mm (8-inch) telescope a value of 12.5.

Telescopes, contrary to general expectations, do not increase the brightnesses of extended objects. Even stars become extended sources when magnifications sufficiently high to enable their seeing disks (see Section 2.5.2) to be resolved are used. The reason why extended objects are not increased in brightness is that the telescope magnifies (Section 2.2) them. The increased amount of radiation gathered by the telescope from the object is therefore spread over a greater area. At best, a telescope will show an extended object with the same surface brightness as when seen with the naked eye. If too high a magnification is used, then the surface brightness of the object will actually reduce when seen through the telescope. Of course the total amount of light received from the object is increased, and this may give the impression that it is brighter. This will particularly be the case for the smaller extended objects because they may be magnified to the point where some of their image spills over into parts of the eye containing the more sensitive rod cells (Section 2.8). Thus, if observing faint galaxies and nebulae is to be your main observing programme, then bigger is not necessarily better. A high-quality efficient telescope, so that as little light as possible is obstructed, scattered or absorbed, and which is used at the optimum magnification of

$$m_{\text{optimum}} = 140\,D \qquad (2.4)$$

where D is the diameter of the telescope objective in metres, is likely to give better results than a bigger but poorer telescope. The latter, though, will still be likely to be better for stellar work where you are observing point sources.

2.5.2 Resolution

The second major function of a telescope is to improve resolution. Resolution is the ability to see close objects separately. The middle star of the "handle" of Ursa Major (see Fig. 2.19, *overleaf*), known as ζ UMa or Mizar, can be seen with the naked eye to have a second star, Alcor, close to it. The separation of Mizar and Alcor is about 11.5′, and most people's eyes can resolve two such objects when their separation is down to 3 to 6′. However, Mizar itself is not a single star, but two stars separated by 14.5″. This is too close for the stars to be separated by the unaided eye,

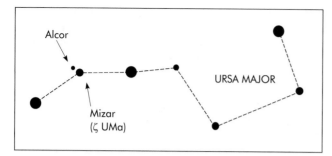

Figure 2.19 Mizar and Alcor (separation exaggerated for clarity).

and so they are below the resolution of the eye. Even a 50 mm (2-inch) telescope however will be able to show the two stars separately. The brighter component of Mizar is yet another double star, but this time with a separation of only 0.04″. In the absence of the Earth's atmosphere this could be resolved using a 3 m telescope, but in practice it cannot be separated by any existing telescope today. The star is known to be double, not through being resolved, but because the two stars orbit each other, and the Doppler shifts[5] of their spectral lines reveal the nature of the system.

The resolution of a telescope of diameter D metres, for two stars are of similar brightness, is given in seconds of arc by

$$A = \frac{0.12}{D} \qquad (2.5)$$

If the stars are of very different brightnesses, then to see them separately, a much larger telescope will normally be needed than that given by Eq. (2.5). Thus Sirius (α CMa) is a double star with a maximum separation of 11″. But the brighter star (Sirius A) is some 10 000 times brighter than the fainter (Sirius B). A 0.3 m (12-inch) to 0.5 m (20-inch) telescope is therefore needed to resolve this pair rather than the predicted 50 mm (2-inch) telescope.

Equation (2.5) suggests that if you are the owner of a 0.3 m (12-inch) telescope, then you should be able to

resolve stars separated by only 0.4″. If you are lucky, then perhaps once or twice a year you will be able to reach this limit. Most of the time, though, the turbulence of the atmosphere (Section 2.9) will limit the resolution to 1″ or 2″, or worse. Telescopes larger than 0.3 m will rarely, if ever, reach their limiting resolution. Only telescopes of 0.15 m (6-inches) or less will regularly achieve their theoretical resolution.

As you become a practised observer, you will find that you will be able to use your telescope at its resolution limit more and more frequently. This is because, although nights when the atmospheric turbulence is calm enough to allow continuous resolution down to 0.5″ are very rare, there are moments of clear-seeing on even the worst nights. Thus, with practice, you can ignore 10 or 20 minutes' worth of observing when the turbulence was bad, and note all the details of the object in the 1 or 2 second clear-spells that happen now and then. This is an even more important technique with planetary observations, and explains why a skilled observer can often see more detail than can be photographed or imaged with a CCD detector. It is also possible to take steps to reduce the atmospheric turbulence, as discussed in Section 2.9.

In order to use a telescope visually at its resolution limit, the images must be sufficiently magnified that they are separated, when seen in the eyepiece, by more than the eye's resolution limit. For an observer with averagely acute vision, the minimum magnification for this is then

Minimum magnification = 1300 D $\qquad (2.6)$

where D is the telescope's diameter in metres. People with less acute vision will need to use even higher magnifications.

2.6 Cleaning and Aluminising

The optical surfaces of lenses and mirrors are shaped very precisely (to within 50 or 60 nm – about two millionths of an inch), and it is very easy to ruin them. In most cases they will also have a thin coating – a reflecting layer for a mirror, and an anti-reflection layer for a lens – that may easily be damaged. Nonetheless, the optics of a telescope must be kept clean, or light will be absorbed and scattered, leading to poorer images.

So how do you clean optical surfaces? The first absolutely wrong thing to do is to breathe over the surface

---

[5] Waves, such as light or sound, will be observed to have a longer wavelength when the emitting object is moving away from us compared with when it is at rest with respect to us. Similarly, the wavelength will be shortened when the object is moving towards us. This change in wavelength is known as the *Doppler shift*.

and whip out a handkerchief to polish it with. The second absolutely wrong thing is to take lenses or eyepieces apart in order to clean their inner surfaces. If inner surfaces of lenses, eyepieces or of the optics in enclosed designs such as Schmidt–Cassegrains need cleaning, there is really no alternative but to take them back to the supplier. With care however, it is possible to clean the external surfaces of optical components.

Let us take lenses first. No liquids (except isopropanol, which is advertised for the purpose in astronomy magazines) should ever be used to clean lenses unless you are absolutely sure that they do not possess an anti-reflection coating. Even then, liquids are best avoided because they can percolate between components and become impossible to remove, or they may damage the glue in cemented lenses. A clean stream of air or gas may be used to blow dust off the surface. A hair-drier, set to cold, and with a clean well-washed handkerchief over the air intake to act as a filter, can be used for this. Alternatively, it is possible to purchase from some electronic component suppliers, aerosol cans containing an inert gas under pressure for the same purpose. Care needs to be taken when using the latter however, because liquid propellant can be expelled for the first second or two of a burst. You should never blow dust off with your breath, tempting though it is to do this; you are certain to deposit some drops of saliva on to the surface and mark it. Once dust etc. has been removed, the surface may be *very gently* wiped with the cleansing tissue sold for the purpose by most opticians (do not use an ordinary paper handkerchief).

The same comments as those just outlined for lenses, generally apply to cleaning mirrors, with one exception. The exception is when the mirror's surface has had a protective coating applied. The protective coating is usually of silicon monoxide, and is applied immediately after the mirror has been aluminised. It prolongs the life of the reflective coating, and is sufficiently hard to withstand very gentle washing. With such a mirror, therefore, it is possible to wash it – just sluice water over the surface, never rub the surface with anything. Distilled water with a drop of detergent is all that should be used for this. Follow the wash by several rinses in pure distilled water, and then stand the mirror on edge in a draught-free spot to dry.

The lenses in eyepieces are just as delicate as those of an objective lens. Yet many observers casually stuff eyepieces into their pockets when not in use. At the very least, eyepieces treated in this way will pick up fluff and dirt. However, if they cohabit the pocket with the sweet wrappers, pens, spare film cassettes, other eyepieces etc. such as are customarily carried by the average observer,

then they will quickly become scratched, lose their anti-reflection coatings, and generally be ruined for any purpose except the least critical work. Keeping spare eyepieces on your person however does have the advantage that they are kept warm and so are less inclined to dew-up (see Section 2.7) when brought out for use. Thus the ideal solution is to put your eyepieces into a clean sealed container before putting them into a pocket. This can be a bit of a nuisance when you are changing eyepieces frequently, but given the cost of eyepieces, it is well worth the effort. Most eyepieces are now supplied in plastic containers which are adequate for their protection when they are not in use. If not, then you will probably find something suitable in a hardware store; small sealable plastic containers are sold there for use on picnics etc. You can use soft bags to keep eyepieces in if they are not kept with other hard objects. Thus plastic bags may be used (though these tend to develop holes quickly) or bags made from lint-free cloth. The latter may be obtained by putting a handkerchief or tea towel through the washing machine a dozen or more times.

Eventually mirrors, or more rarely lenses, will need to have their coatings replaced. The usual reflective coating on mirrors is aluminium and this is deposited on to the surface in a vacuum chamber. There are several specialist firms which undertake this work and they usually advertise regularly in the popular astronomy magazines (Appendix 2). It is possible to apply a silver coating to a mirror at home, though it will need renewing every month or two. Details of the process may be found in chemistry books where it is called the "silver mirror test". If the anti-reflection coatings on lenses need renewing, they will need to be returned to the original manufacturers, though with care, the coatings on lenses should last for many years.

2.7 Dewing-up

The cleanest and most perfectly produced telescope will quickly be reduced to ineffectiveness through condensation forming on the surfaces of its optics. Dewing-up affects all telescopes, but is especially severe with refractors and Schmidt–Cassegrain telescopes which have lenses at the outer end of the instrument.

The best solution to dewing-up is to stop it happening to start with. Thus a telescope in an enclosed observatory will be much less prone to the problem. In the open, or on nights when the temperature drops quickly in an atmosphere with a high humidity, dew will form on exposed optical components rapidly, sometimes within a few minutes. The onset of dewing-up can be delayed, or even prevented, through the use of a dew-cap (see Fig. 2.20). This is simply an open tube projecting beyond the end of the telescope. The longer it is, the more effectively it will reduce dew. The inside of the dew-cap should be painted matt black or covered with black paper to avoid reflections when viewing bright objects like the Moon. A very low-power heater (a few watts) placed inside the dew-cap can eliminate the problem of dewing-up entirely, but may cause a slight degradation of the images through the production of convection currents near the telescope's objective.

Reflectors frequently have an open skeleton as their telescope "tube". Their mirrors are then as likely as the front lenses of refractors to dew-up. The solution is the same as for refractors – the skeleton tube should be enclosed. This can be done very simply on a temporary basis with aluminium foil wrapped around the tube and held in place with tape etc. (though this can introduce annoying reflections). If you find that dewing-up is a persistent problem with your telescope, then it is worth permanently blocking-off the gaps in the skeleton tube. This can easily be done with thin plywood or hardboard, painted matt black on the inside and cut to fit into the gaps.

If, despite all your precautions, your telescope has dewed-up, what can be done? There are two possibilities: give up observing for the night, or remove the condensation. If you opt for the latter, then NEVER wipe the surfaces with a cloth etc. to get rid of the moisture. The only method of getting rid of the condensation without risking damaging the optics is to evaporate it with a stream of clean, warm air. This is likely to reduce the quality of the subsequent images because of the introduction of convec-

tion currents, but since the alternative is to see nothing, it is definitely the lesser of the two evils. A small hair-drier used on the lowest setting will quickly rid the optics of their condensation. WARNING – if your telescope is dewing-up, then conditions will be damp. If you use a mains-powered hair-drier, then it MUST be connected to the power via an Earth-Leakage Circuit Breaker (sometimes known as an RCCB, RCD, or a Rapid Circuit Breaker). Far better, however, is to purchase one of the low-voltage driers sold for use when camping or caravaning, and to use it from a battery supply. In bad conditions you may find you have to clear the condensation every few minutes.

The greatest source of condensation is you – it is remarkably annoying if you check your telescope optics for dewing-up, find them clear, but then go and breathe over them to leave a misty surface. Eyepieces (and spectacles) are especially prone to becoming misted through your breath. If eyepieces are kept on your person when not in use, then they will be slightly warm and less inclined to dew-up (but see warnings in Section 2.6). If they do dew-up, then they should be cleaned using the hair-drier as above. The temptation to wipe them with a handkerchief (or even a finger) should be strictly resisted.

2.8 Observing Techniques

An experienced observer will see far more through a telescope than a novice, simply because he or she knows how to look. The experience is not difficult to acquire, and comes under the heading of observing techniques. It is a collection of largely *ad hoc* pieces of advice and practice.

2.8.1 Dark Adaption

It is a matter of common experience that upon walking out of a brightly lit room into a dark night, one is effectively blind. After a few minutes, however, some things can be discerned, and after half-an-hour it is likely that you will be able to see well enough to move around quite confidently. This improvement in vision is due to the phenomenon of *dark adaption*. Two factors contribute to the increased sensitivity of the eye after a time in the dark. The first, and best known, is that the pupil of the eye dilates from a diameter of 2 to 3 mm to one of about

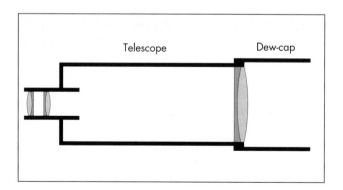

Figure 2.20   A dew-cap in use.

7 mm. This increases the amount of light entering the eye by a factor of 7 or 8.

By far the largest effect, however, is an improvement in the sensitivity of the eye's retina. This improves eyesight by a factor approaching 100. The improvement arises through the structure of the retina. This contains two types of detecting elements, called (from their shapes) *rod cells* and *cone cells*. The cone cells come in three varieties and detect light of three different wavelengths, giving us colour vision. They are however comparatively insensitive. The rod cells are of only one variety, but are much more sensitive than the cone cells. In normal, brightly illuminated conditions, the light-sensitive element of the rod cells (a molecule called rhodopsin or visual purple) is completely used up so that our vision is almost entirely due to the cone cells. When we go into dark conditions, the rhodopsin regenerates, taking about 20 to 30 minutes to become completely restored. As it regenerates, so the sensitivity of the rod cells improves, reaching its maximum after half-an-hour or so. Hence after 30 minutes or so in the dark, our vision has improved in sensitivity by a factor of several hundred. But our vision is then monochromatic because there is only the one type of rod cell (hence the expression "all cats are grey in the dark").

It takes half-an-hour or so for your eyes to become dark-adapted. But it only takes a second or two of exposure to a bright light (such as looking down a torch beam) to destroy that dark adaption. Preservation of dark adaption, by avoiding bright lights, is therefore a priority for an observer, especially when observing the fainter objects. Yet, it will often be necessary to consult a star atlas or catalogue or to see a sheet of paper on which one is making notes or drawings, while in the midst of an observing session. How can loss of dark adaption then be avoided? The answer lies in the sensitivity ranges of the rod and cone cells in the retina. Rod cells are sensitive out to a long-wavelength limit of about 600 nm (orange-red). The red-variety of cone cell, however, is sensitive out to about 700 nm (very deep red). Thus if we use light in the range 600–700 nm, it should be possible to see items, such as star catalogues, using the red-sensitive cone cells, while at the same time not destroying the dark adaption of the rod cells. Hence it has become the practice for observatories, when in use, to be illuminated by red lighting, and to use red LEDs (Light Emitting Diodes) as indicators on equipment. Most such lights and filters, however, fail in their purpose because they have short-wavelength leaks which allow through light to be picked up by the rod cells. Thus a really deep red safelight sold for use when processing blue-sensitive photographic emulsions (NOT a safelight for use

with photographic printing paper) is the only suitable illumination that will preserve dark adaption completely.

Most astronomical observing is undertaken with dark-adapted eyes, and this leads to what is often a major disappointment for beginners. When you look at pictures of interstellar nebulae and galaxies in books, they are usually magnificent spectacles in glowing blues, greens and reds. But when you see them directly, they are too faint to trigger our colour vision and so appear as wispy greyish hazes (Chapters 8 and 9). Even the brightest examples, such as the Orion nebula, are likely to be seen only as a greenish glow. With time and experience, the disappointment vanishes, and the thrill of your first detection of a faint and difficult planetary nebula or galaxy more than compensates for the lack of colours.

2.8.2 Averted Vision

A phenomenon which is closely related in its causes to dark adaption is that of averted vision. It is that a faint object may be difficult or impossible to see when you try to look at it directly, but if you look slightly to one side of where it should be, then it will suddenly flash into view quite clearly. When you look directly at an object, the image falls on to a part of the retina (called the *fovea centralis*) that is very rich in cone cells and which has very few rod cells. It is therefore a region of low sensitivity in the retina (though it does give the best colour vision and resolution). By looking to one side of the object in which you are interested (averting your sight slightly), its image is forced to fall on to a part of the retina away from the *fovea centralis*. Since the proportion of rod cells to cone cells increases with distance from the *fovea centralis*, the image is therefore falling on to a more sensitive part of the retina and thus becomes much easier to see.

Averted vision is not an easy technique to use initially. As soon as the object becomes visible, it is an almost automatic physiological reaction to look at it directly, when of course it disappears again. With practice, however, you will find that you can continue to look to the side of the object, while observing it in detail. Though it takes some acquiring, the trick of observing by averted vision is well worth mastering, and will greatly extend the range of objects which you can see through your telescope.

For really faint objects, you may find that "nudging" the telescope slightly, either by hand or with its slow-motions, may enable the object to be seen while it is in motion across the field of view, even though it disappears again when stationary.

2.8.3 Seeing

With any good-quality telescope larger than 75 mm or so, the details that you can observe of an object in the sky will largely be limited by the turbulence of the Earth's atmosphere. This turbulence, or seeing, is discussed further in the next section. Here we are just concerned with how best to observe in its presence.

If you look at a star through your telescope using as high a magnification as possible, then you will see the turbulence of the atmosphere affecting the image in several ways. The image will be "dancing" around; moving irregularly around its mean position, it will be brightening and fading, and if you have a high enough magnification to see it as a disk,[6] then its shape will be changing. If you look at an extended object such as a planet, then each point on its surface will be behaving like the image of the star. The main effect of atmospheric turbulence on extended objects is thus to reduce the contrast in the image and for the finer details to be lost.

However, turbulence by its nature is very variable. Occasionally, therefore, there will be moments when the column of the atmosphere through which the light is passing in order to reach your telescope briefly becomes quiescent. When this happens, the image will suddenly be seen in all the detail that the telescope is capable of providing. The observing technique to cope with the atmospheric turbulence is thus just to wait for these fleeting clear-spells, and then to note the details of the object, ignoring the blurred image normally to be seen.[7] The frequency of moments when the telescope performs at its

diffraction limit (Eq. (2.5)) decreases rapidly as the size of the telescope increases. Thus a 75 mm (3-inch) telescope will reach its diffraction limit most of the time, a 0.15 m (6-inch) periodically on most clear nights, a 0.3 m (12-inch) telescope on a few good nights a year, and a 0.5 m (20-inch) telescope probably never. With the larger size telescopes, the image will be better on some occasions than others even if not reaching the diffraction limit, and so the same observing technique still applies.

2.8.4 Finding

Finding charts and the details of how to set your telescope on to specific objects are dealt with later in this book. But what of the general principles of the process? One of the most important accessories for your telescope is a good finder telescope. With the telescopes at the cheaper end of the market, the finder telescope is sometimes rather shoddy or even non-existent. In such cases it is worth replacing the finder supplied by the manufacturer with a better quality instrument.

A good finder telescope should be a short focal length, 50 mm or 75 mm (2 to 3-inch) diameter refractor, with a low-power, wide-angle eyepiece. The eyepiece should also have illuminated cross-wires, with the level of illumination capable of being varied. The finder telescope should be attached to the main telescope via a mounting that holds it firmly but without inducing any stresses, and which may be simply adjusted to bring the two instruments into mutual alignment.

A widely used design for mounting the finder on to the main telescope is a pair of circles (see Fig. 2.21). The setting screws on opposite sides of each circle may be adjusted[8] to move the finder in any direction with respect to the main telescope. To align the finder, the main telescope is set on to a bright object (just by pointing and sweeping until you come across it), and the setting screws (or other adjustment) on the finder's mounting adjusted until the finder's cross-wires are centred on the same object. The adjustments should then be locked and the finder will be aligned permanently with the main telescope.

With such a finder telescope, it will be easy to observe all but the faintest objects. The main telescope is simply

---

[6] This is not, of course, the real disk of the star, but the "seeing" disk, that is the spreading-out of the (almost) point source which is the star, through the combined effects of diffraction in the telescope, aberrations of the telescope optics, and the effects of the atmosphere. It will normally be 1″ or 2″, compared with the true disk of the star, which will usually be much less than 0.01″.

---

[7] When observing in this manner it is important to see what is there and not what you expect. You should approach each observing session with the intent to observe what is there, and not be influenced even by what you saw on previous occasions. Preconceptions (perhaps from looking at spacecraft images) may easily lead you to force an interpretation on to these ephemeral details which is not true. As a salutary example of this we have the Martian canals, which once the suggestion had been made that you should be seeing lines on Mars, were then "observed" by many people.

---

[8] It is good practice to have the finder separately attached to the main telescope by a safety cord, in case the finder slips out of its circles while you have slackened the mounting screws to adjust its alignment.

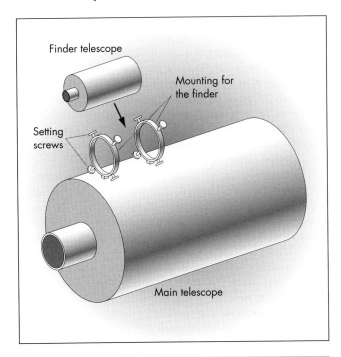

Figure 2.21 A mounting for a finder telescope.

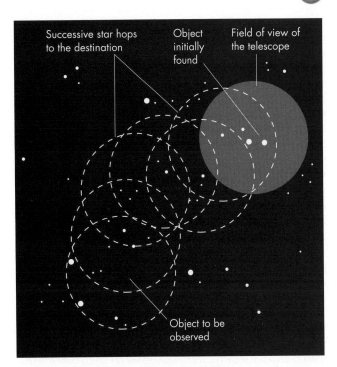

Figure 2.22 Star hopping.

pointed in roughly the right direction by eye, or set on to the coordinates of the object. That object should then be visible in the finder telescope, or become so with a little sweeping around the area. Once visible in the finder, the telescope is moved until the object is centred on the cross-wires. It should then be visible in the main telescope, especially if a low-power, wide-angle eyepiece is used, and it can be centred and observed as wished.

For objects too faint to be seen in the finder, or if your telescope has a less than ideal finder, then another approach must be adopted. This is called *star hopping* and is essentially the same process as learning your way around the constellations that was discussed in Chapter 1, but on a smaller scale. You will need a reasonable star atlas (Appendix 2), because you will need to prepare finder charts (Section 2.10). The principle behind star hopping is just to find a known object and then move from that object (hop) via others that you can recognise until you reach the object of interest (see Fig. 2.22). A low-power, wide-angle eyepiece (Section 2.2) is best for this purpose, and you should know how wide a field of view it gives with your telescope. You start by finding (using the finder telescope, or just by pointing by eye and sweeping) a known object in the main telescope. This will usually be

a nearby bright star or easily recognisable nebula. You centre this object in the field of view, and identify the first feature on the way to your destination. This feature may just be another star, but a group of stars in a recognisable pattern is better, so that you can be sure of the identification. You move the telescope to centre this feature and then search for the second feature. The process is repeated until you find the object that you want. Note that it is very easy to get confused and lost when star hopping; it is amazing how many times you seem to recognise a set of stars etc. quite clearly, only for it to turn out to be the wrong one. If this happens, just go back to your starting point and try again. With experience, you may be able to dispense with a specially prepared finder chart and work directly from the star atlas, but a finder chart will help to begin with.

2.8.5 Guiding

If you intend to obtain images from your telescope by photography or with a CCD camera which will require exposures longer than a few seconds, then you will normally need to guide the telescope. This is because, even though your telescope is driven to follow the motions of objects in the sky (and you must have a driven telescope for exposures longer than about a tenth of a second), the drive is unlikely to be perfect. The imperfections in the drive will cause the image to move during the exposure and produce a trailed or blurred record (see Fig. 2.23). Even if the drive were perfect, there would still be occasions when guiding would be necessary. Thus with very long exposures, the changing amount of refraction by the Earth's atmosphere will lead to the objects in the sky moving at a non-uniform rate, or you may wish to image an object that itself is moving across the sky, such as a comet or an asteroid, and therefore need to guide the telescope on that object to get a sharp image.

The simplest means of guiding a telescope uses a guide telescope attached to the main instrument. This second telescope will need to be of a reasonable size and with a focal length comparable with that of the main telescope. Thus the finder telescope (Section 2.8.4) is not usually adequate to act as a guide telescope. The guide telescope needs to be equipped with a high-magnification illuminated cross-wire eyepiece. The observer then centres the object to be imaged on to the cross-wires, and uses the telescope slow-motions to keep it centred throughout the exposure. Guiding is a skill which it takes some practice to acquire. Your first few attempts will probably be worse than if you had not guided at all! However, perseverance will soon pay off with pin-sharp images.

Sometimes you may want to image an object which is too faint to see in the guide telescope, or you may want to image a galaxy or nebula which is too diffuse to provide a good aiming point. In these circumstances, the technique of offset guiding can be used. The main telescope is set on to the object to be imaged. The mounting of the guide telescope is then adjusted *without moving the main telescope*, until the guide telescope points to a nearby star which is bright enough to see. Guiding then continues using this star.

If you do not have a guide telescope, then sometimes the main telescope can be used to obtain the image and to guide at the same time. It is possible to obtain adapters for attaching the camera to the telescope which have a tiny mirror or prism to divert a small part of the field of view to a separate eyepiece. After the telescope has been aligned on to the object to be imaged, the mirror or prism is

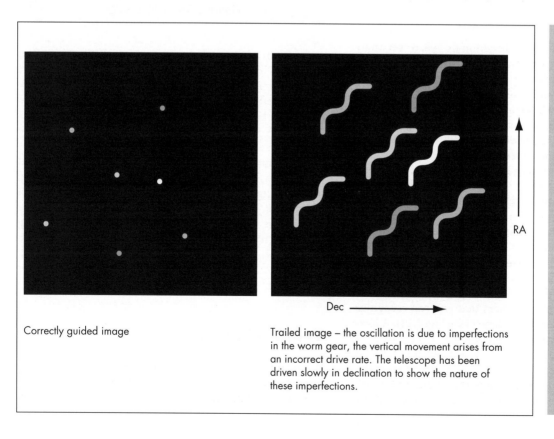

Correctly guided image

Trailed image – the oscillation is due to imperfections in the worm gear, the vertical movement arises from an incorrect drive rate. The telescope has been driven slowly in declination to show the nature of these imperfections.

Figure 2.23 The effect of an imperfect drive upon an unguided exposure.

adjusted until another object can be seen in the eyepiece, and that is then used for guiding purposes (see Fig. 2.24). Alternatively a partially reflecting mirror can be used. This would be set at 45° to the optical axis and reflect about 90 per cent of the light into the camera, while the remainder passes straight through into the guiding eyepiece. A variation upon this theme is now to be found with some CCD cameras. These have a main imaging chip and a smaller second CCD chip alongside it. The main chip is used to obtain the desired image, while the second chip is aligned with another object. The image from the second chip is read out at frequent intervals during the main exposure, and used to guide the telescope automatically.

Some CCD cameras provide another way of overcoming tracking errors, since it is easy to add their images together within a computer. If the drive of the telescope is good enough to give sharp images for a short exposure, then many such images can be obtained, lined up, and added to give the total required exposure. This is a procedure that is used on many telescopes, including the Hubble space telescope, whose exposures are of limited duration because of changing stresses as it passes through the Earth's magnetic field.

Computers can also play a part in reducing or eliminating the need for guiding. With the sophisticated computer-controlled Schmidt–Cassegrain telescopes now available (Table 2.1), the error in the drive (see Fig. 2.23) may be determined and that information fed into the computer. The computer then controls the motor speeds in order to compensate for the physical deficiencies of the drive mechanism. Such computers can also calculate and allow for the effects of refraction in the Earth's atmosphere, so that guiding becomes almost unnecessary.

If your telescope is on an alt–azimuth mounting (Section 2.4) then there is an additional complication: the field of view rotates as the telescope follows an object across the sky. This field rotation cannot be corrected by guiding. For short exposures, the blurring effect may not be noticeable, but for longer exposures, a *field de-rotator* will be needed. This is a device which rotates the camera at a suitable rate to compensate for the rotation of the field of view. A high level of mechanical precision is needed in its construction, and its rate of rotation will need to vary as the telescope points at different parts of the sky. It is therefore not an easy device to make for yourself. The manufacturers of computer-controlled alt–azimuth-mounted telescopes will normally be able to supply a suitable field de-rotator for their own instruments, but these do add significantly to the cost. For mountings such as the Dobsonian (see Fig. 2.18), the whole mounting and telescope may be placed on to an inclined-plane drive (see Fig. 2.25) to provide equatorial-type tracking for a period of a few minutes.

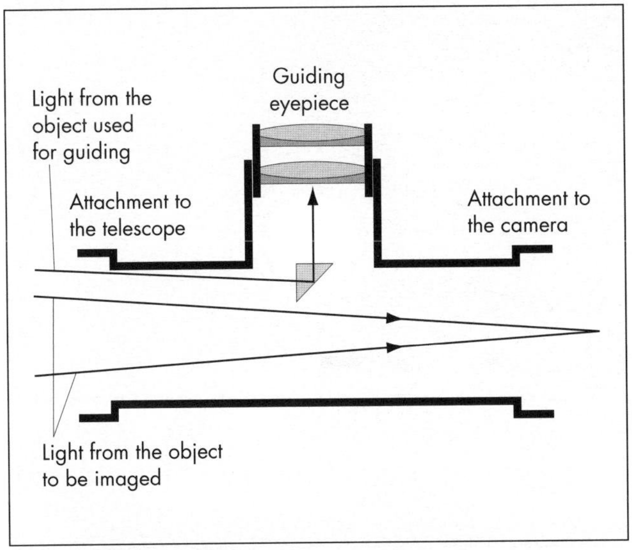

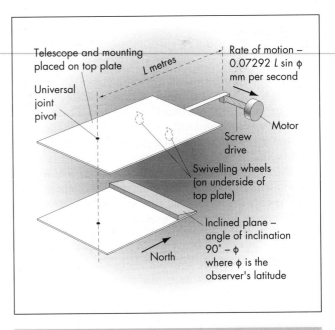

Figure 2.24 Cross-section through a camera adapter with an offset guiding eyepiece.

Figure 2.25 Semi-exploded view of an inclined plane equatorial-drive platform for a Dobsonian-type alt–azimuth mounting.

2.8.6 Apodisation

Apodisation is a paradoxical technique, for it requires that in order to improve your telescope's performance you seem to be taking actions which will make it worse. Indeed, in general, apodisation will make your telescope's performance worse, although in some crucial area it will improve it. For some purposes, that improvement can outweigh the other disadvantages. Apodisation consists of deliberately altering the shape and/or size of the image that the telescope produces for a point source such as a star; this is often called the *instrumental profile or point spread function* of the telescope. The changes that you may make to your telescope under this heading are not permanent, and are usually made only for the purposes of one type of observation. There are several ways in which telescope performance can be improved through apodisation, and these are best studied by a few examples.

Resolution

The normal diffraction-limited resolution of a telescope is given by Eq. (2.5). But this can be improved significantly, to

$$A = \frac{0.08}{D} \qquad\qquad (2.7)$$

by covering over the centre of the objective. Only a thin annulus, a centimetre or so in thickness of the outer rim of the objective, should be left uncovered. This does, of course, reduce the intensity of the image very considerably, but if you are looking at bright objects, the improvement in the image can be well worth it. There is also a considerable increase in the relative brightness of the diffraction fringes. The reason for the improvement in resolution is easy to see if we imagine the lens or mirror to be made up from a series of annuli. Each annulus will have its resolution given by Eq. (2.7). But when we combine the low resolutions of the small-diameter annuli near the centre of the objective with the better resolutions of the larger annuli near its edge, the average becomes that given by Eq. (2.5). This type of apodisation therefore just consists of eliminating the low-resolution parts of the objective.

Both Eqs (2.5) and (2.7) apply when the two objects to be resolved are of similar luminosity. If they are very different in brightness, then usually the resolution which can

be achieved will be much worse than these two equations imply. One reason for this is that diffraction leads to the images of point sources having faint circular fringes around their central cores (see Fig. 2.26). The image of a faint object near to a bright one may therefore be masked by the fringes from the bright object. By covering the objective with a star-shaped mask (see Fig. 2.27), bright diffraction spikes are produced (see Fig. 2.28) but the circular fringes disappear. The mask can be rotated until the faint object lies between two of the diffraction spikes, and it can then be seen free from the effects of the fringes. In this way Sirius' companion, which is a white dwarf some 10 000 times fainter than the main star, can be found with quite a small telescope.

Contrast

When observing extended objects which have a low contrast, such as many of the planets, the diffraction spikes and fringes (see Figs 2.26 and 2.28) have the effect of reducing the contrast in the image. Features, even if theoretically resolved, thus become impossible to distinguish from each other. Both fringes and spikes can be

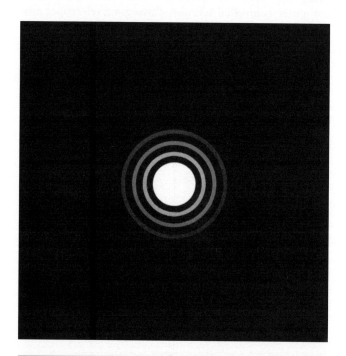

Figure 2.26 Diffraction fringes around the image of a star.

Figure 2.27 A star-shaped mask.

Figure 2.28 The image produced when a star-shaped mask is used over a telescope objective, showing the diffraction spikes and the elimination of the diffraction fringes.

eliminated for telescopes which do not have support arms for a secondary mirror (that is, refractors and most types of Schmidt–Cassegrains) by another variation on apodisation. In theory, this requires the transparency of the objective to vary smoothly from 100 per cent at the centre to 0 per cent at the edge (ideally, the variation should be Gaussian in shape). This will also have the effect of worsening the theoretical resolution (Eq. (2.5)) by a factor of 2, but this is unimportant when it is lack of contrast which is limiting observations. A variable-density filter could be used to achieve the variation in transparency, but it would be impossibly expensive. In practice, therefore, a much cruder and cheaper approach is used which still allows a big improvement in contrast to be obtained. A set of masks for the objective is made out of narrow-mesh chicken wire. These have an outer diameter equal to that of the objective, and a central hole. The central hole varies in size from a centimetre or two to nearly equal to the objective's diameter. The masks are all overlaid on to the objective, and provide a rough approximation to the required variation in transmission.

2.9 Twinkling

A beautiful clear frosty night in winter, you look up and see the stars as brilliant pinpoints twinkling away against a velvety black sky – a perfect night to get the telescope out and do some serious observing right? Wrong! – almost certainly when you do look through your telescope on such a night the images will be poor. The twinkling, which is also known as *scintillation*, is a variation in the brightness of the star arising from turbulence high in the Earth's atmosphere. It combines with the seeing, which is variations in the image structure and position arising from turbulence in the lower atmosphere, to cause the image to be blurred. It is the combination of scintillation and seeing that limits the resolution (Section 2.5.2) of telescopes larger than a few tenths of a metre to one or two seconds of arc even under good conditions.

Atmospheric turbulence changes in a time of a few milliseconds. Thus if we obtain an image of a star with an exposure of a millisecond or so, the atmospheric effects will be frozen, and we can see their effect upon the image in detail. What we see is perhaps surprising; the image consists of a large number of small spots covering an area a second or two of arc across (see Fig. 2.29). The size of the turbulent zones in the atmosphere is typically a few tens of centimetres across. Telescopes smaller than this can thus often be used at their diffraction limit, when they are then completely covered by a single such zone. Larger telescopes however will always have light arriving at their objectives which has passed through several zones. Each of the spots in the instantaneous image is a part of an interference pattern produced when the light from two zones combines at the focus of the telescope.[9] As the zones change, the spots making up the image come and go and move around, so that the image we detect on an exposure of a tenth of a second or longer is the combination of many such spots covering an area much larger than the theoretical resolution limit of the telescope.

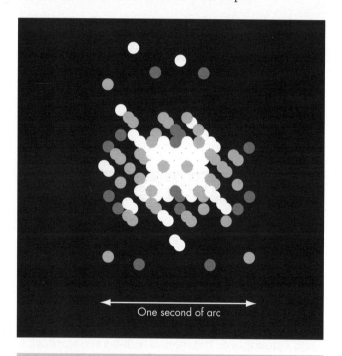

One second of arc

Figure 2.29 The highly magnified image of a bright star obtained with an exposure of a millisecond or so using a large telescope.

[9] These spots (or speckles) can be used to obtain images at or near the resolution limit of the telescope. The technique is called *speckle interferometry*. It is a complex process and the interested reader is referred to the bibliography for further details (Appendix 2).

Can anything be done about scintillation and seeing? Scintillation results from turbulence high in the atmosphere, and therefore little can be done to reduce its effects, except to choose nights when it is at a minimum. If a cold frosty night with stars twinkling away is a poor choice, what conditions are likely to give minimum scintillation? The answer will vary with the location of the observer, but frequently the best nights are when there is little wind, and just a suggestion of a mist forming. The stars will not appear so bright under these conditions, but their images will often be better than on apparently much clearer nights. These are, however, also the conditions under which dewing-up (Section 2.7) is likely to be a problem.

Seeing is the result of low-altitude turbulence, with much of it arising in the immediate vicinity of the telescope. If you have some choice in your observing site, even if only moving from one part of the garden to another, you may find that you can reduce the effects of seeing by siting the telescope in an area of lower than average turbulence. Such a best position is likely to change as the wind direction and speed vary, and so it will need a good deal of experimenting to find the optimum position for all conditions. With a fixed or a portable telescope, seeing is likely to be improved by the use of wind shields. These are just large panels of fine mesh (not solid panels, as these are more likely to cause turbulence than to reduce it), which you can position around the telescope to shield it from the wind. Experimentation will again be needed to find the best positions for such panels. As an additional bonus, wind shields will protect the telescope from the direct effects of the wind which otherwise can cause an unprotected instrument to shake so much that observation becomes impossible.

2.10 Finder Charts

Faint objects will usually need to be found using your main telescope. The process of star hopping from a prominent object to the one that you want has already been discussed (Section 2.8.4). For this process you will often need to prepare a finder chart showing the stars and other objects that you can expect to see on the way to the object of your search. You will also need to have a finder chart for observing the asteroids and the planet Pluto, because these will look like stars. Over a period of time, they will move against the background stars, but it can be hours or days before that motion is perceptible. A finder chart will therefore be needed so that you can identify which of the star-like images in your field of view is actually the asteroid or planet.

The preparation of a finder chart is straightforward, but does require the use of a good star atlas or catalogue (Appendix 2). You need to know the size of the field of view of your telescope, and the faintest stars that you are likely to be able to see. Remember that this will vary depending on the sky background. If you are observing with a bright Moon in the sky or during twilight, you will be able to see far fewer stars than on dark clear nights. From the star atlas or catalogue you draw out the region around the object of your interest. It is best to do this on tracing paper, or on the clear plastic sheets sold for use on overhead projectors (you will also need a suitable pen to write on OHP sheets, obtainable at good stationery suppliers). You can then view your finder chart from either side and orient it to align with the field of view. You should only mark on your finder chart those objects that you can actually expect to see. A circle of the size of the field of view through the eyepiece that you are using and which can be superimposed upon the finder chart is also a useful accessory.

Often the popular astronomy magazines (Appendix 2) will contain finder charts for objects of current interest such as novae and comets, and these can be very helpful to the observer. For less ephemeral objects, such as galaxies, planetary nebulae, Messier objects, NGC objects etc., ready-prepared finder charts may be available in specialised books (Appendix 2).

2.11 Keeping a Log Book

We all like to gawp at the more spectacular sights to be found in the night sky, and there is nothing wrong with that. However, if your observations become nothing more than gawping, then they are soon likely to lose their savour. Science, whether conducted by large teams of professionals using multi-billion pound devices, or by a lone amateur with a home-made telescope, relies upon the accurate recording of results, and upon the ability of others to duplicate and verify those results, for its successes. Thus you should get into the habit of recording the details of your observations at an early stage in your work.

It is possible to record your results on odd scraps of paper etc.,[10] but these are liable to become lost or to get out of order. It is far better to have a book in which to keep a record of your observations. Ideally, this should be a quality hard-back note book so that it can survive rough handling and still be readable 50 years after it was used. As a minimum, the following details should be noted down for every observing session:

Observer
Date
Start time (remembering to specify the time zone, daylight saving time etc.)
Finish time
Location
Weather conditions
Seeing and scintillation[11]
Telescope(s) used

---

[10] One eighteenth century French astronomer, who had better remain anonymous, is supposed to have written down his observations on to planks of wood and then planed them clean again when he decided the records were no longer needed!

[11] You should use a numerical scale for this. Either develop your own system based on the smallness and sharpness of the stellar images, or use Pickering's scale which is based on the diffraction rings (see Fig. 2.26):

Seeing	Scale	Image appearance
Very bad	1	Image twice the diameter of the third ring
Very bad	2	Image occasionally twice the diameter of the third ring
Very bad	3	Image about the same diameter as the third ring, and brighter in the centre
Poor	4	Airy disk often visible, portions of diffraction rings sometimes visible
Poor	5	Airy disk always visible, portions of diffraction rings frequently visible
Good	6	Airy disk always visible, portions of diffraction rings always visible
Good	7	Airy disk sometimes sharply defined, rings seen complete or as long arcs
Excellent	8	Airy disk always sharply defined, rings seen complete or as long arcs in motion
Excellent	9	Airy disk always sharply defined, inner ring stationary, outer rings momentarily stationary
Excellent	10	Airy disk always sharply defined, all rings stationary.

Continued on next page

and then for each separate observation:

> The object observed
> The time of observations
> Eyepieces and magnifications used
> Notes on the appearance of and/or a dimensioned
>   sketch of the object
> Comments on any unusual features

If you are obtaining photographs or CCD images then you should also record:

> Details of the camera in use
> Temperature settings (if applicable)
> Exposure time
> Time of the exposure
> Number of combined exposures (if applicable)
> Focus settings
> Guiding details
> Use of filters
> Use of Barlow lenses, eyepiece projection, telecom-
>   pressors etc.
> Any image processing (on CCD images)/details of the
>   chemical processing (for photographic images)
> Any comments (such as on accuracy of guiding etc.)

If this seems a long list, then just reflect that it could be your record of one of your observations that guarantees that a future comet or nova or supernova bears your name for the rest of eternity.

If you are taking part in an organised observing programme, or have an observing programme of your own (such as monitoring variable stars), then you are likely to be provided with a form for recording the essential details by the organisers of the observing programme, or you can devise your own form to minimise the labour involved in keeping the records.

For planetary work, Antoniadi's scale can be used:

1   Perfect seeing, without a quiver
2   Slight undulations in the image with moments of calm
     lasting several seconds
3   Moderate seeing with large tremors
4   Poor seeing with constant troublesome undulations in
     the image
5   Very bad seeing, scarcely allowing even a rough sketch
     to be made.

2.12 Discoveries

If you do suspect that you have found a new comet or nova etc., what then should you do? The first step is to ensure that it is not a well-known object. It would be highly embarrassing to report a bright new comet in Andromeda, only to find that it is actually M31, the Andromeda galaxy! If you have not looked at the sky for a while, then even Jupiter and Venus can appear as unexpected objects. Note also that if you are using an unfiltered CCD camera (Chapter 11), then red stars such as cool supergiants and Mira variables will appear much brighter relative to nearby hotter stars than they do visually or on photographic atlases. With CCD images it is also essential to have more than one image in case the suspect object is actually a flaw or cosmic ray strike etc. on one of the images. "Ghost" or false images can also appear from multiple reflections in the telescope if there is a very bright object just outside the field of view.

In the early stages of your observing you should therefore assume that new-to-you objects are *NOT* startling new discoveries. Later on, however, when you are thoroughly familiar with the sky, it is possible that you may be the first to observe a new comet or nova etc. Observations of any suspected discovery should always be made more than once, with a separation of an hour or more between the observations. If you think you have spotted a comet, but cannot detect any motion, then be very sceptical about your discovery. You should also check a good-quality star catalogue or atlas to ensure that it is not a galaxy, and an up-to-date almanac to ensure that it is not an already known comet. If you are certain of your observations, then to ensure that you are credited with the discovery, you must report it through the correct official channels. Many national astronomical societies (Appendix 1) have procedures set up for this and you should check with them first.

More generally, comets, novae, supernovae, outbursts of unusual variable stars, new features on planetary surfaces, etc. may be reported to the Central Bureau for Astronomical Telegrams (CBAT). Non-urgent, conventional changes in variable stars can be reported to the variable star section of your national astronomical society (Appendix 1). The discovery of a new asteroid, after which you are entitled to name it subject to certain restrictions, such as not using an existing name, is somewhat more complex. First you have to obtain at least three separate positions for the object so that its orbital parameters can be determined. It then has to be retrieved at two successive oppositions before it can be added to the list of known asteroids. Reports should go in the first instance to the Minor Planet Centre (MPC). Meteor fireball reports should go to the International Meteor Organisation (IMO). All these organisations use e-mail as their main form of communication, and their up-to-date details may be found on their home pages on the World-Wide-Web. If you do not have access to e-mail etc., then reports can be made by telephone or fax, and contact numbers may be obtained from your national astronomical society (Appendix 1).

All reports by whatever medium should include the following information:

> Your name
> Your address
> Your contact details (e-mail or telephone/fax number)
> Date and UT time of observation
> Observation method (such as naked eye, visual
>   telescopic observation or telescope CCD
>   observation etc.)
> For new contributors, some background information
>   on your observing experience

and then the details of the object discovered:

> For Comets this should include
>   An accurate position and estimated rate of motion
>   or a series of timed positions
>   A description including total magnitude, size
>   diffuseness, central condensation, tail etc.

> For Supernovae this should include
>   Position and identity of the host galaxy
>   Magnitude of the suspected supernova
>   Its offset in seconds of arc and the direction from
>   the galaxy's nucleus
>   How you have determined that this is a new object
>   (for example, comparison to named atlases,
>   previous observations etc.)
>   Whether you have checked for known supernovae,
>   minor planets etc. in the region

> For Novae this should include
>   Position (as accurately as possible)
>   Magnitude
>   How you have determined that this is a new object
>   (for example comparison to named atlases,
>   previous observations etc.)
>   Whether you have checked minor planets etc. in the
>   region

Finally, do you fancy having a star named after you? No problem – just send your cheque for £100 (or $100 etc.) to "Astro-Scams Ltd" (or to the authors!) and they will name a star in your honour. Of course, no one else will use that as the star's name, but the company has not done anything illegal. Stars, with only a very few exceptions such as Barnard's star, are not named after individuals; they are given numbers or other codes from star catalogues. Thus α Centauri, BD 935 + 14.6, HD 130675, SS433 etc. are all *bona fide* star names; Fred's star is not. Beware therefore of the various confidence tricks which appear with monotonous regularity in what may otherwise be quite reputable sources, and which advertise just such a naming service.

Chapter 3

The Sun

3.1 Warning

Severe damage to the eyes and even blindness can be caused by observing the Sun incorrectly. Never look directly at the Sun either with the naked eye or through binoculars or a telescope, unless correctly using a suitable professionally produced filter (Section 3.2). Smoked glass and over-exposed film (which are sometimes suggested as suitable filters) are NOT safe; they may still let the infrared part of the Sun's radiation through.

It is also possible to damage the telescope or eyepieces if the focused energy from the Sun falls on to part of the telescope structure. This may particularly be a problem with designs like the Schmidt–Cassegrain, Cassegrain and Newtonian where the secondary mirror is close to the focal point of the primary mirror. It may also be a problem with any design of telescope while setting the telescope towards the Sun.

3.2 Observing the Sun

Despite the warning given above, solar observing is perfectly safe provided that suitable precautions and procedures are followed. The Sun is then a rewarding subject for study with its changing appearance as sunspots come and go in their 11-year cycle (see Fig. 3.1 and Fig. 3.2, *overleaf*). It is also a rewarding object in that it is viewed during the daytime and if the atmosphere is clear enough for solar work, then it is almost certainly a very pleasant day to be outside using your telescope.

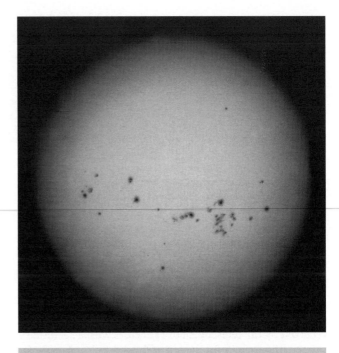

Figure 3.1 An image of the Sun near sunspot maximum showing many sunspots, and with some plages towards the limb. Limb darkening can also clearly be seen.

3.2.1 Stopping-down

In contrast to most other areas of astronomy, there is no problem with lack of light when observing the Sun. Indeed, as already mentioned in Section 3.1, the opposite

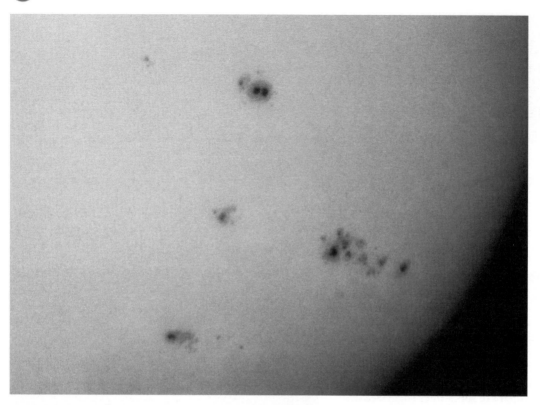

Figure 3.2 A detailed view of some sunspots.

problem of too much energy is ever present, so that there is danger of damage to eyes or to the telescope. For most purposes a 50 mm to 75 mm (2″ to 3″) aperture will be sufficient. Owners of larger telescopes will therefore need to reduce the effective size in order to undertake solar work. This is most simply done by stopping the telescope down. Stopping-down just involves placing an opaque screen of cardboard, thin plywood or metal, which has a hole 50 mm to 75 mm across cut in it, over the telescope's objective. Only the light passing through the hole is then received by the telescope. For reflectors with a secondary mirror, the hole in the screen will need to be placed off-axis so that it is not obscured by the secondary mirror. Care should be taken to ensure that the screen is firmly attached for if it were to fall off or blow away, the full aperture of the telescope would suddenly be gathering sunlight with possibly disastrous results to the observer or instrument.

The need to stop down a telescope before solar observing means that the normal improvement in resolution with increased size does not apply. Although full-aperture filters (see below) do enable the telescope to work at its nominal resolution, this is not normally an advantage.

The reason for this is that it is almost always the Earth's atmosphere that limits resolution, not the telescope. The heat from the Sun causes turbulence that makes daytime seeing generally poorer than that at night. The best time to observe the Sun is normally an hour or two after sunrise. Then the Sun is high enough in the sky to be above the thickest part of the atmosphere, but it has not had time to produce much turbulence. Professional solar observatories go to great lengths to reduce the effects of solar heating, such as being sited in the middle of lakes, or at the tops of tall towers.

3.2.2 Eyepiece Projection

This is the simplest, cheapest and one of the best ways of observing the Sun. First stop your telescope down to an aperture of 50 mm or so, and put in an inexpensive, low-power eyepiece.[1] Then point the telescope at the Sun (Section 3.2.5) without looking through it. Hold a sheet of white cardboard 0.2 to 0.5 m (8″ to 20″) behind the eyepiece. You should then see at least a part of the projected image of the Sun (see Fig. 3.3). Move the telescope until the whole disk of the Sun can be seen, and adjust the eyepiece until the image is focused. Features such as sunspots, limb darkening, and possibly plages and granulation (see Figs 3.1 and 3.2), should then easily be seen.

If you intend observing the Sun frequently, then it is worth constructing a framework to attach to the telescope, to hold the screen on to which the image is projected.

3.2.3 Full-aperture Filters

These are filters which cover the whole aperture of the telescope and which eliminate the infra-red radiation as well as reducing the visual intensity of the Sun. They are made from thin plastic such as Mylar, which is then aluminised. When such a filter is in place it is safe to look

directly through the telescope at the Sun, even with quite large telescopes. Full aperture filters must be purchased from a reliable supplier, it is not possible to make your own. Usually the manufacturer of the telescope will supply a suitable filter as a standard accessory. The filters are quite fragile and need to be handled carefully. If a filter becomes damaged, then it must be discarded, and a new one purchased. When using a full aperture filter do not forget to blank off your finder telescope, or its projected solar image can produce a nasty burn if it falls on to your skin.

3.2.4 Solar Diagonals

These devices are also known as Herschel wedges, and enable the Sun to be seen through a normal eyepiece. They are not suitable for use at apertures greater than 50 mm (2″). They consist of a plain piece of glass set at 45° to the optical axis of the telescope. The plain glass reflects about 5 per cent of the light to the side and into the eyepiece (see Fig. 3.4). They are wedge shaped so that the reflection from the second surface does not interfere with that from the first. A problem with these devices is that 90 per cent of the solar energy passes through them and emerges from the back of the telescope. It is easy to forget this when observing, and only to remember when the smell of scorching can no longer be ignored!

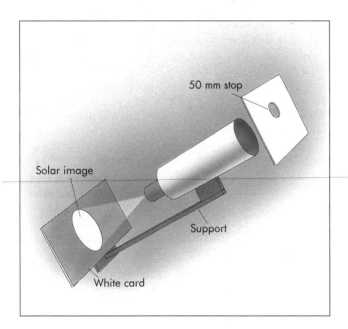

Figure 3.3 Eyepiece projection of the solar image.

---

[1] Even stopped down, a lot of energy will be gathered by the telescope. This could heat the eyepiece to the point where it might be damaged. It is therefore wise not to use an expensive eyepiece for eyepiece projection. Furthermore, the more expensive eyepieces are likely to have more lenses, and to be more complex in construction than simpler eyepieces, therefore they are more likely to be vulnerable to heat damage.

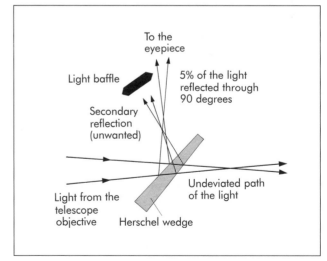

Figure 3.4 The solar diagonal or Herschel wedge.

3.2.5 Finding the Sun

Given that it is obvious where the Sun is in a clear sky, finding it in a telescope might not seem to be a problem. However you cannot look through the finder or the main telescope (unless they both have full aperture filters) in order to set on to the Sun. The simplest approach is therefore to circularise the shadow of the telescope. First point the telescope roughly towards the Sun, and then look at its shadow on the wall, the ground, or on a suitable board placed behind the telescope. Then move the telescope, without looking through it, so that the shadow decreases in size. When the telescope is pointing at the Sun, the shadow will be at its minimum size, and square-on to the telescope will be circular (or whatever may be the shape of the cross-section of the telescope).

3.3 Solar Observing Programmes

The main features to be observed on the Sun are the sunspots (see Figs 3.1 and 3.2). These are regions of the Sun ranging from a few thousand to a few hundred thousand kilometres across wherein magnetic fields lead to a reduction of the surface temperature from about 6000 K to 4000 K. The light emitted from a sunspot region is therefore reduced compared with that from the normal surface of the Sun by about a factor of 5, and so spots appear dark compared with the photosphere. Individual sunspots and sunspot groups may last from a day or two to several months. The larger groups of spots develop and change in complex ways over that time. The number of spots and the positions on the Sun at which they appear change over a period of about 11 years in the well-known sunspot cycle.

The development of individual sunspots and groups may be deduced from regular observations. The period of rotation of the Sun as seen from the Earth varies from about 26.75 days near the solar equator to 31.5 days near the solar poles. Thus only spots and groups with lifetimes greater than about a month are likely to be seen both forming and fading. The development patterns of sunspots must therefore be built up by observing many spots at various stages in their lives.

The total area covered by sunspots is measured by the Zurich sunspot number, R. This is given by

$$R = k\,(10\,g + s) \tag{3.1}$$

where g is the number of sunspot regions, either single spots or groups,

s is the number of individual spots, both single spots and those within groups,

k is a personal correction factor.

The personal correction factor is to take account of differing instruments, observing conditions etc. between observers. It has to be determined by comparing your estimates of R over a period of time with the officially recognised values. Once your personal value of k has been found, then you can make regular estimates of the sunspot number and follow the solar cycle. Many national astronomical societies (Appendix 1) have sections for solar observers. Joining such a section will enable you to contribute your observations to the world-wide monitoring of solar activity.

The Sun does not rotate as a solid body, but varies its rotation period from the equator to the poles as mentioned above. The various rotation periods may also vary with time. Sunspots do move by small amounts over the surface of the Sun, but these movements are sufficiently small that they may be ignored and so sunspots may be used to determine the rotation of the Sun. An interesting observing programme therefore, which is again suitable for collaboration with others, is to monitor the solar rotation rates by measuring the movement of sunspots across the solar disk. Remember that in converting from a motion across the disk to a rotation period, you will need to allow for the projection of the three-dimensional shape of the Sun on to the two-dimensional visible disk, for any inclination of the solar equator to the line of sight, and for the Earth's motion around its orbit. Data for making these conversions may be found in the *Astronomical Ephemeris* (Appendix 2).

Other features of the Sun which you may be able to observe include solar flares, granulation, plages and the limb darkening.

Between two and five solar eclipses occur every year. But unlike lunar eclipses (Chapters 4 and 10), they are only visible from a small part of the Earth's surface. Partial and annular eclipses are observed in the same way as the uneclipsed Sun. The total solar eclipse, when the solar chromosphere and corona may be observed, is only visible over a very tiny portion of the Earth. Normally, therefore, if you wish to observe a total solar eclipse you will need to travel to where it is occurring. This is becoming an increasingly popular pastime, and several travel firms advertise package tours in the astronomy magazines (Appendix 2) whenever a solar eclipse is due. The naked eye, binoculars or a telescope may be used for observing a total eclipse without the need to use filters etc. However, all the normal solar-observing precautions must be used until the last part of the solar photosphere has disappeared, and care must be taken not to be looking through the telescope when the photosphere reappears at the end of the eclipse.

3.4 More Advanced Work

So much light is available from the Sun that there are numerous special devices and instruments to enable observations of it to be made that are impossible for other objects. Some of these devices can be constructed by a competent DIY enthusiast, others may be available commercially at affordable prices. Many however are very complex and expensive, and are likely therefore to remain the concern only of the professional solar observer. A brief summary of some of these devices is given below, the interested reader however will need to refer to sources in Appendix 2 for further information.

3.4.1 The Prominence Spectroscope

This is a device to enable solar prominences (see Fig. 3.5) to be observed. Like the H-α filter (see below), it is able to reveal the prominences because in the red light of hydrogen at 656 nm they are as bright as the solar photosphere. The spectroscope uses a transmission grating, and the entrance slit is aligned with the limb of the Sun. A second slit at the focus of the spectroscope then isolates the hydrogen line. Observations are made through this second slit with a normal eyepiece. The telescope should be stopped down to 50 mm (2″) or so, but a filter is not necessary because the light is spread out into the spectrum. Prominence spectroscopes can be made by the DIY enthusiast, but they are neither easy to construct nor to use.

3.4.2 The H-α Filter

This device also enables prominences to be observed, but in addition allows features on the disk of the Sun such as flares, filaments and chromospheric granulation to be seen (see Fig. 3.6). It is a very narrow-band filter centred on the hydrogen line at 656 nm. It is not suitable for home construction, but is available commercially for about the same price as a 0.2 m (8″) Schmidt–Cassegrain telescope.

3.4.3 The Spectrohelioscope

This is yet another device for observing the Sun in the light of the red hydrogen and other spectrum lines. It is just about possible for a very good and well-equipped DIY

Figure 3.5 Solar prominences.

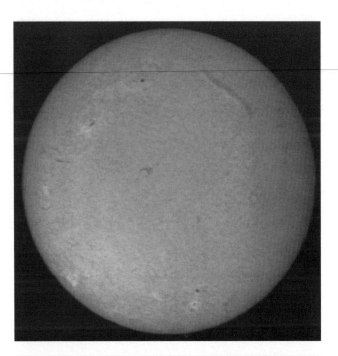

Figure 3.6 An H-α image of the Sun (sometimes called a spectroheliogram).

enthusiast to construct a spectrohelioscope. They are not available commercially. The spectrohelioscope is essentially the same as the prominence spectroscope except that the entrance slit of the spectroscope is oscillated back and forth to cover the whole disk of the Sun. The second slit therefore has to move concurrently with the first slit in order to remain centred on the spectrum line. The Sun is observed through the second slit, and if the oscillations are rapid enough, an image of the whole Sun in the light of the H-α or any other spectrum line will be built up.

3.4.4 The Coronagraph

The concept of the coronagraph is simple, getting one to work is diabolically difficult. In a coronagraph an artificial eclipse is produced by blocking out the image of the photosphere of the Sun with an opaque disk, so that the solar corona may be observed. The problem with the instru-

ment is that the least amount of scattered light from the solar photosphere will overwhelm the faint glow of the corona. The optics of the coronagraph have therefore to be of very pure glass, with no bubbles or striae, they must be very clean and dust-free, and the instrument must be placed at high altitude on a site with a very clear and transparent atmosphere.

3.4.5 Solar Spectroscopy

It is quite within the ability of a DIY enthusiast to construct a spectroscope (Chapter 11) that will suffice to reveal a wealth of information about the Sun. Solar spectroscopy can be used to identify the elements present in the Sun, to measure its rotational velocity from the Doppler shifts of spectrum lines, and to detect the intense magnetic fields present in sunspots through the broadening or splitting of spectrum lines via the Zeeman effect.

4.1 Introduction

The Moon is at one and the same time one of the easiest and one of the most difficult objects to observe. When you are starting in observational astronomy it is one of the easiest and amply rewarding of objects; it is simple to find in the sky and in the telescope. Even the most basic of telescopes will show some details, and those details will vary over the month as the Moon's phases change. Later you will find it a most difficult and frustrating object to observe if you decide to specialise in lunar observation. This is because there is such a wealth of detail to be seen, but much of it is only visible when the shadows are exactly right and under the best observing conditions. Some tiny craters may thus only be available for a few hours once a year – and if luck runs true to form, that will be when you have sent your mirrors away for re-aluminising! Though much detail is visible even in the simplest instruments, the finest work requires the highest quality instruments, with squeaky-clean optics, and a design such as a refractor that minimises diffracted light. The use of an apodisation mask (Section 2.8.6) may sometimes be advantageous. Furthermore, despite the spacecraft that have visited and orbited the Moon, and the centuries of observations by terrestrial observers, it is not exhausted as a subject for original work. Transient Lunar Phenomena (TLPs – Section 4.5), for example, are almost exclusively studied by amateur astronomers.

However let us leave those more esoteric considerations for the moment and start with the basics.

4.2 Naked-eye Work and Binoculars

The Moon is about half a degree across as seen from the Earth, and is therefore large enough for details to be observed without the use of a telescope at all. The changing phases are easily seen. The phases are due to the differing amount of the illuminated part of the Moon that we can see, and not to the Earth's shadow falling on to the Moon, as is often mistakenly thought (see Fig. 4.1, *overleaf*). The latter effect produces lunar eclipses (see below). Some lighter and darker patches can also be seen and, when the Moon is near to full, produce the "Man-in-the-Moon" illusion (see Fig. 4.2, *overleaf*). When the Moon is a very thin crescent, light reflected from the Earth illuminates the dark part sufficiently for it to be seen faintly from the Earth. We then get the "Old-Moon-in-the-New-Moon's-arms", with the whole disk of the Moon visible faintly and with one edge a brilliantly lit crescent (see Fig. 4.3, *overleaf*).

Lunar eclipses, when the Moon passes into the Earth's shadow (see Fig. 4.4, *overleaf*), can also be observed with the naked eye (see Fig. 4.5, *overleaf*). The total eclipse phase lasts for an hour or two. Little additional detail over that to be seen with the naked eye will be seen through a telescope, because the Moon is full when an eclipse occurs. The term *total eclipse* is rather misleading because the Moon rarely blacks out completely. Light is scattered in the Earth's atmosphere to illuminate the darkened parts of the Moon. Often this scattered light has

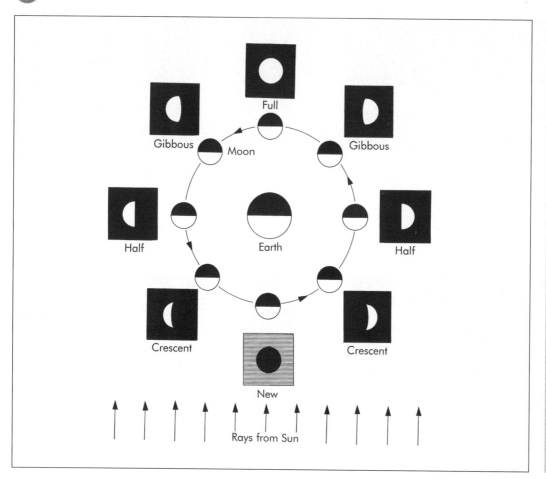

Figure 4.1 The phases of the Moon.

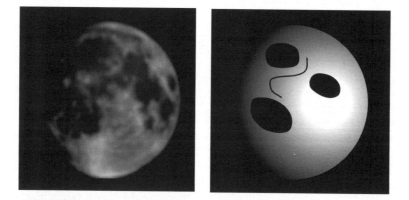

Figure 4.2 The Man in the Moon (the orientation here is as seen in the sky and not inverted as seen through a telescope).

Figure 4.3 The Old Moon in the New Moon's arms.

Figure 4.5 The Moon partially eclipsed (*top*) and totally eclipsed (*bottom*).

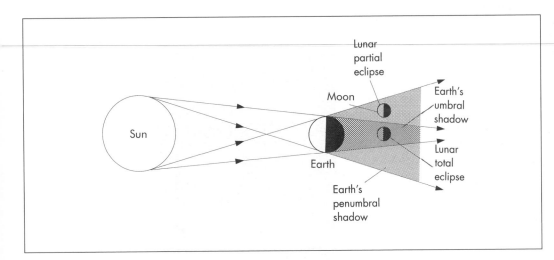

Figure 4.4 The geometry of a lunar-eclipse.

a reddish tinge, as in Fig. 4.5. If a major volcanic eruption, however, has thrown a lot of fine dust high into the atmosphere, the total eclipse phase can be almost black. Lunar eclipses are rarer than solar eclipses, but because they are visible from an entire hemisphere of the Earth, they seem from a single observing site to be more common. Depending on the alignments of the Earth, Moon and Sun, there can be from zero to three lunar eclipses in a year. Predictions of their timings and visibility will be found in astronomical almanacs and yearbooks, the astronomy magazines (Appendix 2), and close to the event, in newspapers and on radio and television.

The Moon moves across the sky quite rapidly; its average rate of motion is about half a second of arc per second of time. It thus moves through its own diameter (of half a degree) in about an hour. This is fast enough to be seen easily with the naked eye. If the Moon is close to a bright star, then its alignment should be noted, or better still sketched. Repeating the observation after an hour or two will show a change in the Moon's position with respect to the star. If there are no convenient bright stars close to the Moon, then pairs of more distant ones can be used. By selecting a pair of stars whose connecting line passes through or close to the Moon, then the Moon's position can be plotted on a suitable star map. Again, a repeated observation after a short interval will show the Moon's motion. A piece of string is a useful aid here. It can be stretched at arm's length and aligned on the stars to improve the accuracy of the estimated position of the Moon. Continuing this process over a month will enable the Moon's path across the sky to be plotted on to a star map.

A much more ambitious project would be to plot out the long-term path of the Moon across the sky. The Moon's orbit is inclined to the Earth's orbit around the Sun by just over 5°. If you do plot out the Moon's monthly motion as just described, you should therefore find that it oscillates by 5° either side of the ecliptic.[1] However, the plane of the Moon's orbit slowly rotates in space because of tidal interactions with the Earth. It takes 18.61 years to complete the circuit. Hence, over that time, you should find that the Moon's path across the sky takes it through all points in a band ±5° either side of the ecliptic.

Trying to see the very thin crescent of the newest or oldest possible Moon is an exacting naked-eye task. The brilliance of the Sun means that the Moon is lost in the scattered solar radiation and so is not visible immediately before or after new Moon. (Be very careful not to look at the Sun itself if you are searching for the Moon while the Sun is still in the sky – see Chapter 3.) You can congratulate yourself if you can find the Moon within 30 hours of new Moon, and if you do better than 20 hours then you are getting on very well indeed. The earliest records of

sighting the crescent Moon are about 15 hours from conjunction.[2]

A pair of reasonable binoculars with ×7 or ×10 magnification will show more details of the Moon, but will only just show the larger craters (see Figs 4.6 and 4.7). The maria and terrae (dark and light sections respectively) will be seen far better than with the naked eye. It will also be possible to notice the effects of libration (see later in this chapter) with binoculars. Notice the changing distance from the right-hand lunar limb of the prominent circular maria, Mare Crisium, resulting from the differing orientations of the Moon between Figs 4.6 and 4.7. It is possible to obtain binoculars with magnifications higher than ×10, and very good they can be. They will, however, only be usable if they have a stand or other means of support; it is almost impossible to hold binoculars by hand at magnifications higher than ×10 for more than a minute or two.

4.3 The Moon through the Telescope

Even a small telescope will show a wealth of detail, including craters, mountain ranges, valleys, etc. when pointed at the Moon (see Fig. 4.8). A magnification ranging from

Figure 4.6 The crescent Moon as seen through binoculars.

---

[1] The *ecliptic* is the Sun's path across the sky over a period of a year. It is usually marked on star maps and star globes, and is inclined to the equator by 23.5°. Since the Sun's movement in the heavens is actually due to the Earth's orbital motion around it, the ecliptic is also the plane of the Earth's orbit, if we imagine it extended out into distant space.

---

[2] *Conjunction* is when an object is in line with the Sun. Strictly, for the Moon therefore it only occurs when there is a solar eclipse. However, the term is also used rather more loosely to mean the point when the Moon is nearest the Sun in the sky.

Figure 4.7 The nearly full Moon as seen through binoculars.

about ×50 up to whatever the available eyepieces and the atmospheric conditions will allow can be used. At the higher magnifications, using the slow-motions on the telescope mounting (if it possesses them) can give the feeling of flying over the surface. Furthermore, the shadows near the terminator (the line dividing the Sunlit part of the Moon from the dark part) move sufficiently quickly that changes can be seen in an hour or two. From one night to the next, the whole scene can change. Most detail will be seen near the terminator, where the shadows are strongest. Those regions of the Moon which have the Sun high in their sky will appear comparatively featureless. Indeed, at full Moon, it may be quite difficult to see craters at all, though other features such as the crater rays are then at their most prominent.

Unless you intend to become a dedicated lunar specialist, there is little point in learning the names of all the features that you can see. Nonetheless, becoming familiar with the major features such as the maria, and larger craters is a good way of honing your observing skills. A good map of the Moon (Appendix 2) is all that you need for this, though make sure that it is one produced for astronomers with south at the top. Many maps produced by the space agencies such as NASA have north at the

Figure 4.8 a and **b**. Two views of the Moon as seen through a reasonable telescope (oriented as seen in an astronomical telescope with south at the top).

top and are almost impossible to use when looking through a telescope. Sketching the changing appearance of a prominent feature, such as the crater Plato, the

Alpine Valley or Straight Wall, will not only be good practice but will also highlight the enormous changes in the appearance of lunar features with different phases of the Moon.

A good test of your observing skill, which you can repeat over the years to see how it is improving, is to determine the smallest craters that you can see. You will need a good map of the Moon for this, but you should only use it after you have completed your observations. It is very easy to imagine that features are visible if you know that they are there, and so it is best to make your observations in an unprejudiced manner, and to check them out afterwards. Small features will only be visible when the shadows are right (see Fig. 4.9), and are generally most prominent in the region of the terminator. Under conditions of good seeing, with the atmosphere limiting resolution to one second of arc, features as small as 0.5 km across should be discernible.

Figure 4.9 A chain of small craters near the large crater Copernicus which provide a good test of observing skills.

4.4 An Optimum Telescope for Lunar Work

If you decide to progress beyond simple lunar observing, then you need to consider the best type of telescope to use, and the best ways of using it for this purpose. Unlike observing stars (Chapter 7), the size of the telescope is much less important than its design when it comes to looking at the Moon. The reasons for this are quite complex, but are important enough to consider in some detail. They lie in the fashion that contrast and resolution are affected by the instrumental profile (Sections 2.5.2 and 2.8.6). The instrumental profile is the way in which the telescope produces an image of a point source like a star. For a high-quality telescope there will be a sharply defined bright central spot, called the Airy disk, and surrounding diffraction fringes (see Fig. 2.26). When the telescope has arms supporting a secondary mirror, diffraction at their edges will add spikes to the diffraction fringes (see Fig. 4.10). If the optics of the telescope are not of the highest quality,

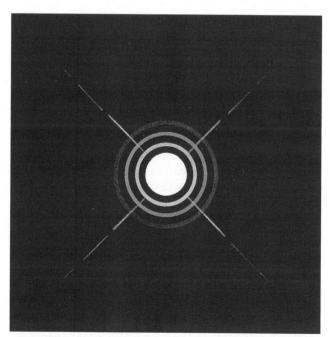

Figure 4.10 The instrumental profile for a telescope with a secondary mirror supported on arms from the sides of the telescope tube.

then the image of a star will be far worse than these diffraction-limited cases.

Now the resolution of a telescope is normally taken to be given by Eq. (2.5), and this will generally be true for two close stellar images of roughly equal brightnesses. That equation, however, is based only on the size of the central Airy disk. The sizes and brightnesses of the diffraction fringes and spikes were not taken into account when it was derived. The fringes and spikes, however, cannot be ignored for extended objects like the Moon (also the Sun, planets, galaxies and nebulae). This is because although, at any given point within the instrumental profile, the fringes and spikes are faint compared with the central peak, they cover a much larger area. The total energy going into the fringes and spikes can therefore be quite large compared with that in the Airy disk. Modern large telescopes are often specified as having x per cent of the energy going into the Airy disk, where x is around 60 per cent to 80 per cent. That means, of course, that 20 per cent to 40 per cent of the energy gathered by the telescope is going into the diffraction fringes and spikes.

Now, if we consider a single point being observed on the surface of the Moon, this point will be the equivalent of a star, and the light from it will be spread into the instrumental profile. This will reduce the intensity of the point being observed. Much more importantly, however, is that the point we are observing is surrounded, in the case of an extended object, by a myriad of other points of roughly similar brightnesses. Each of those points will have its light spread into the same instrumental profile. The light in the fringes and spikes from these surrounding points will thus overlap into the region of the point we are actually observing. This does not affect the nominal resolution, but it does reduce the contrast between areas of differing brightness very considerably. The actual resolution will therefore be impaired because of the increased difficulty of distinguishing between two adjacent areas of differing brightnesses.

To take an example, consider a shadow inside a lunar crater. The shadow may actually have a brightness of, say, 5 in some arbitrary units, while the surrounding sunlit areas have a brightness of 100 in the same units. The actual contrast between the shadowed and the sunlit areas is thus 1:20. Now consider observing that crater with a telescope that concentrates 60 per cent of the light into the Airy disk. The shadow will have an observed intensity of 3 units plus the light diffracted into it from the surrounding areas. The diffracted light intensity will be 40 units provided that the instrumental profile is smaller than the area covered by the sunlit areas of the Moon. The observed shadow intensity will thus be about 43 units.

The observed sunlit areas' intensity will remain at about 100, because the light lost from each point into the fringes will be replaced by light from fringes from other nearby points. The observed contrast is thus reduced in comparison with the actual contrast by nearly a factor of nine to 1:2.3. If the resolution is limited by poor atmospheric conditions then the situation can be even worse. With the Moon, it is unlikely to reach the point where no detail at all can be distinguished, but for lower contrast objects like the planets (Chapter 5) there will be many occasions when all that can be seen will be a featureless blob.

It is thus the instrumental profile which primarily determines the quality of images for the Moon and similar objects, rather than the diameter of the objective. In seeking an ideal telescope for lunar observing, a design which produces the cleanest instrumental profile is therefore to be preferred over sheer size. For many purposes, this means a refractor. A good-quality refractor scores on several points over a reflector; there is no central obstruction caused by a secondary mirror, there are no diffraction spikes produced by the support arms for the secondary mirror (some designs of reflector, such as the Schmidt–Cassegrain and the Makzutov, lack these support arms as well), and the reflecting coatings on mirrors can have minute irregularities which scatter a proportion of the light into the outer parts of the instrumental profile, thus increasing the intensity of the fringes relative to that of the Airy disk.

Whatever type of telescope you may be using, the optics should be as clean as possible. Even a small amount of dust on the surfaces of a lens or mirror will increase the scattered light considerably. Great care is therefore needed in cleaning lenses and mirrors, as discussed in Section 2.6.

The technique of apodisation (Section 2.8.6) using a filter which varies in its transmission from 100 per cent in the centre to 0 per cent at the edge of the objective can be useful. This has the effect of eliminating the fringes of the instrumental profile. The central disk is increased in size (see Fig. 4.11, *overleaf*), so that the theoretical resolution is worsened, but this is not important when it is the reduced contrast that is limiting your observations.

4.5 More Advanced Investigations

One area of considerable interest, and which possibly affords scope for original work, is the observation of Transient Lunar Phenomena (TLPs). For several decades some observers have claimed to see occasional temporary

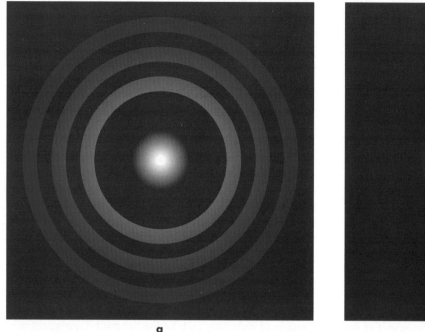

a   b

Figure 4.11 The effect upon the instrumental profile (the image of a star) of apodising using a filter which varies its transmission from 100 per cent at the centre to 0 per cent at the edge of the objective. **a** Normal instrumental profile **b** Instrumental profile with the apodising filter in place.

changes in some lunar features. The events are concentrated in a few areas such as the craters Alphonsus and Aristarchus. A typical TLP consists of the veiling of part of the crater, often the central mountain range, so that it appears less clear than normal. Very occasionally, light emissions may accompany the veiling. The observations are difficult because the changes are slight, and the changing illumination of the Moon causes objects to vary in appearance anyway. So the reality of TLPs is not universally accepted. Nonetheless, if they exist, they could be due to temporary dust clouds produced by gases trapped below the lunar surface escaping and carrying dust from the surface with them. An apparent correlation of TLPs with times of maximum stress in the lunar surface produced by the terrestrial tides, lends credence to this idea. If you are interested in working on TLPs, then you should contact your national astronomy society (Appendix 1) for further information.

Libration has already been mentioned and is shown in Figs 4.6 and 4.7 – it is the phenomenon whereby we are able to see rather more than half the Moon's surface from the Earth. It is often said that the Moon always keeps the same face towards the Earth. If this were strictly true, then only 50 per cent of the Moon's surface could ever be seen. Now, although the Moon's orbital period and rotational period are identical (which is what is implied by "keeping the same face towards the Earth"), the rate of rotation is constant, but the rate of motion around the orbit varies slightly. Thus sometimes the Moon in its orbital motion "gains" a little on its rotation and we can see slightly around one edge, while at other times it "loses" and we see slightly around the other edge. In all, about 59 per cent of the Moon is potentially visible from the Earth. Studying these areas normally on the far side of the Moon is an exacting task because of the very limited times when they are available, but one which has a fascination entirely of its own.

A phenomenon somewhat related to libration is that of *parallax*. The Moon is close enough (385 000 km) for slightly different views to be obtained from different parts of the Earth. The difference is small but adds to the effects of libration. More importantly though, it provides a means of measuring the distance of the Moon from the Earth. If two observers separated by 100 km or so measure the position of the Moon with respect to a close star as accurately as possible and at the same instant, then the distance can be obtained using simple trigonometry. If the separation of the observers is large, then the curvature of the Earth will need to be taken into account. It is also possible for one observer to determine parallax by observing the Moon's position at moonrise and moonset, because the rotation of the Earth will have moved the observer a considerable distance in that time. However to do this, the Moon's motion across the sky, which is much larger than the parallax motion, will need to be offset.

The depths and heights of craters, mountains etc. on the Moon can be determined from the lengths of their shadows. You will need some means to measure these lengths, and an eyepiece with a movable cross-wire (known as a micrometer eyepiece) is the most satisfactory way of doing this visually. Alternatively, the shadow lengths can be measured on photographic or CCD images (Chapter 11). You will also need to know the exact phase of the Moon from a good ephemeris (Appendix 2), so that you can find the altitude of the Sun as seen from the point on the Moon that you are observing. Some quite complex trigonometry is involved in this process, which is beyond the scope of this book, and the reader is referred to sources in Appendix 2 for further information.

Finally, the ages of parts of the lunar surface can be estimated by crater counting. The principle behind this is that craters are being produced all the time by meteorite impacts. The older a surface is, therefore, the more craters it should have. It is difficult to get precise age determinations, but by counting the number of craters in the 4–10 km size range over an area 100 km square, the graph in Fig. 4.12 may be used to get a reasonable estimate.

Photography is discussed in more detail in Chapter 11. It is quite easy to take lunar photographs, because it is so bright that exposures of well under a second are adequate. It is, however, surprisingly difficult to take photographs (or CCD images) showing the same amount of detail as

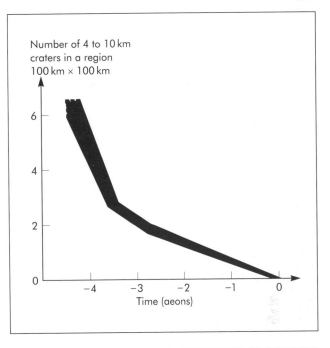

Figure 4.12 The way in which number of craters in an area of the Moon 100 km times 100 km varies with time (an aeon is one thousand million years).

that visible to the eye. This is because, with even a small amount of practice, the eye and brain in combination build up a composite image from the fleeting fractions of a second when the atmosphere is still enough to enable crisp images to be seen. A photographic exposure will rarely be at such an instant, however, and even when by chance it is, the image is unlikely to be equally clear over its whole area.

Chapter 5

The Planets and Minor Solar System Objects

5.1 Introduction

Seven planets – Mercury, Venus, Mars, Jupiter, Saturn, Uranus and Neptune – are angularly large enough to be seen as disks in small telescopes. Of these, Uranus and Neptune at best are only 4 and 2.5 seconds of arc across respectively, and so little or no detail can ever be seen on them. The four Galilean satellites of Jupiter – Io, Europa, Callisto and Ganymede – can just be seen to be non-point sources under the best observing conditions. The remaining tens of thousands of objects (Pluto, the Asteroids, Planetary Satellites etc.) appear star-like at all times. Thus observing the planets and other solar system objects involves similar procedures to observing the Moon (Chapter 4) in the case of the five planets out to Saturn, and to observing stars (Chapter 7) for all the remainder.

The planets and other objects change their positions in the sky because their own motions through space are combined with that of the Earth. These movements can be quite complex (see Fig. 5.1). If you are familiar with the constellations (Chapter 1), then the brightest planets can easily be picked out in the sky. For other objects, however, you will need their current positions in order to be able to find them. The fainter planets and some of the brighter asteroids have their positions listed in various accessible sources such as the astronomy magazines, year books of astronomy, the *Astronomical Almanac* etc. (Appendix 2). With setting circles on your telescope, you can then set directly on to the object. Without setting circles, you will need to plot the position on a star map and then star hop (Chapter 2) to the right spot. Fainter asteroids may just have their orbital parameters listed rather than a proper

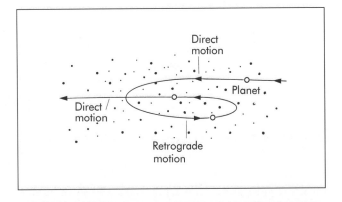

Figure 5.1 The motion of a planet across the sky (schematic).

ephemeris. For these you will then need to calculate their positions from their orbital parameters. There are several computer programs advertised in the astronomy magazines which will perform this calculation. After the brightest hundred or so asteroids, however, you will need to go to the technical literature, or to join a specialist section of an astronomical society in order to find data on them.

We observe from the surface of the moving Earth; this means that the intervals between appearances of a planet in the sky are quite different from the planet's orbital period. The best times to observe a planet are around opposition (see Fig. 5.2, *overleaf*) for a planet further from the Sun than the Earth, and around greatest elongation for a planet closer to the Sun. The interval between successive oppositions or greatest elongations etc. is called the *synodic period* of the planet. The synodic period

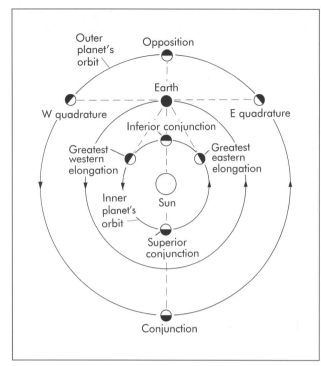

Figure 5.2 Positions of planets with respect to the Earth.

Table 5.1.	Synodic periods for selected objects
Object	Synodic period (years)
Mercury	0.32
Venus	1.60
Mars	2.14
Asteroids	1.2 to 5
Jupiter	1.09
Saturn	1.04
Uranus	1.012
Neptune	1.006
Pluto	1.004

5.2 Mercury, Venus, Mars, Jupiter and Saturn

Techniques for observing these planets have a lot in common with observing the Moon (Chapter 4). In particular, features on them are often of low contrast and so every effort must be made to preserve that detail. The optimum telescope for lunar observation (Section 4.4) will also be the best one for planetary work. Even at opposition or inferior conjunction the angular sizes of the planets are quite small, ranging from 11″ for Mercury to 61″ for Venus. Away from these points they will be significantly smaller. Magnifications up to ×500 or ×600 can therefore usefully be used when the observing conditions permit.

If you have setting circles on your telescope, then it is often possible to find the brighter planets during the daytime. It is even possible to see Venus at its brightest with the naked eye during the day. Seeing planets during the day is very good for impressing visitors, but the reduction in contrast due to the background light will mean that few, if any, features will be seen. The phases of Mercury and Venus (see below), however, should be detectable. In order to find a planet, or bright star, in daylight, the telescope must be in focus before you start searching. If it is in the slightest degree out-of-focus, then the object will blur and be masked by the background light. You will therefore need to leave the telescope in focus from a previous night-time observing session, or to mark the focus position on the eyepiece holder in some way. Some instruments have calibrated focusing mounts, so that all you need do is to note the reading for each of your eyepieces when it is in focus.

therefore governs how frequently you are likely to be able to observe a particular planet. As you may see from Table 5.1, the outer planets, which move very slowly, are visible every year, but much longer intervals occur between opportunities to observe Mars, Venus and some of the asteroids close to the Earth, at their best.

Since all the major planets except Pluto have now been visited by spacecraft, which have provided vastly more detailed and accurate pictures of them than we can obtain from the Earth, the question might be asked "Why bother to observe them at all?". There are two replies to this question. The first is that, for most astronomers, seeing something directly is vastly more satisfying than looking second-hand at someone else's images. The second is that spacecraft have visited the planets for only brief intervals, in some cases just a flyby lasting a few hours. With the possible exception of Mercury, changes are occurring on the planets all the time on time scales ranging from minutes to years. The spacecraft provide a highly detailed but instantaneous snapshot of conditions on the planets. To find what is happening now we need to continue to observe them from the Earth.

Care needs to be taken when observing planets so as not to see imaginary features. When there are many fine details on an object, some of low contrast or just too small to be individually resolved, it is easy to misinterpret what you are seeing. This is not done deliberately, but arises because vision is the result of complex interactions between the eye and brain, and is not just the equivalent of a simple camera. A prime example of this process was the Martian canals, which although non-existent, were actually to be seen only by the most acute observers. Such observers were almost able to see the craters and volcanoes that we now know to be present on Mars. However, the observer's brain would group several of these spot-like features, which could hardly be seen at all and then only fleetingly in moments of utmost atmospheric stability, together and interpret them as lines.

Pre-conceived ideas also need to be avoided. Thus for many years Mercury was expected to be tidally locked on to the Sun. It would therefore always keep the same face towards the Sun, like the Moon keeps the same face towards the Earth. The rotational period of Mercury would thus be the same as its orbital period at 88 days. Observations of Mercury seemed to bear this out. In fact, Mercury's rotational period is 59 days, two-thirds of its orbital period. Since the rotational period is a simple fraction of the orbital period, the same features would re-appear every two orbits. Those features are faint and ill defined, and so observations seemed for a long time to confirm the tidally locked theory.

5.2.1 Mercury

Mercury is one of the most difficult planets to observe, because it is always close to the Sun in the sky, and therefore always to be seen against a bright background. Since it is an inner planet, we have only its dark side visible when it is closest to Earth (inferior conjunction). When we can see the whole of the illuminated side of Mercury, at superior conjunction, it is furthest away from Earth and only about 5″ across. It is also then very close to the Sun in the sky. The best time for observing Mercury is thus around greatest elongation. Since Mercury's orbit is quite elliptical, the elongation (the angular distance of the planet away from the Sun in the sky) can then vary from 18° to 28°. Mercury oscillates from one side of the Sun to the other as we see it in the sky. When it is to the east of the Sun we see it in the evening, and when it is to the west we see it in the early morning.

Mercury can be seen with the naked eye, but it is not an easy object. Copernicus, for example, is reputed never to have seen it at all. It is probably best seen after having found it in the telescope; you can then look along the side of the telescope and so know exactly where in the sky it is to be found.

There are few features discernible on Mercury. At best it is about like looking at the Moon with the naked eye. A few low-contrast light and dark patches are all that are likely to be visible. We know of course from images from the spacecraft *Mariner 10*, that it is rather like our own Moon. It is covered in craters, and has maria, though these are fewer and generally smaller than those on the Moon. None of these features are visible from the Earth through any telescope.

What should be visible, even using a small telescope, are the changing phases of Mercury (see Fig. 5.3). These arise because the planet is closer to the Sun than the Earth, and so we are able to see it from all angles as it goes around its orbit (see Fig. 5.4, *overleaf*).

Occasionally, one may observe Mercury in transit across the disk of the Sun. In making such an observation, it is essential that all the precautions for solar observing (Chapter 3) are followed. A transit occurs when Mercury at inferior conjunction also happens to be near one of the nodes of its orbit. It is then directly on the line between the Earth and the Sun, and can be seen as a black dot silhouetted against the Sun. Transits of Mercury occur at intervals ranging from a few years to a few tens of years on or about 7 May or 9 November. The next transits will be in 2003 (May), 2006 (November) and 2016 (May). Observations of transits used to be important as a means of determining the distance of the Earth from the Sun, and for trying to detect any Mercurian atmosphere. Nowadays, they are primarily observed for their own intrinsic interest.

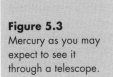

Figure 5.3
Mercury as you may expect to see it through a telescope.

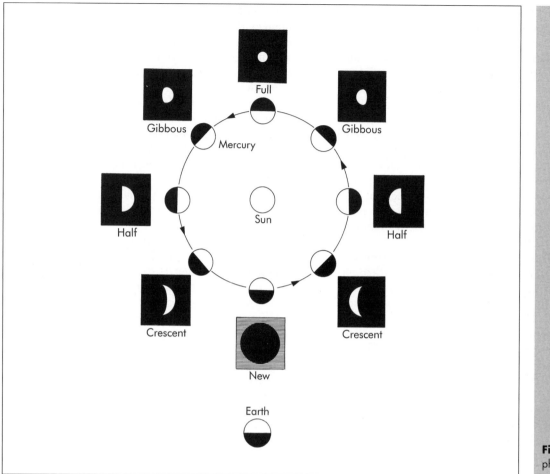

Figure 5.4 The phases of Mercury.

5.2.2 Venus

Venus is an inner planet like Mercury, and therefore many of the same comments apply. Its greatest elongation[1] is about 47°, so that much of the time it may be observed against a black background. It goes through phases, and since Venus is bigger and brighter than Mercury these are easily observed (see Fig. 5.5). That is however about all that is easily observed of Venus; despite being at times the closest object to the Earth after the Moon, very little detail can ever be discerned. This is partly because like Mercury, when Venus is at inferior conjunction and closest to Earth, we can only see its dark side. More importantly, however, is the immensely thick atmosphere which forever shrouds the surface and hides it from optical observation. Radar can penetrate Venus' atmosphere and reveals a world with plains, huge volcanoes and mountain ranges, but also with many less familiar features, but none of these can be seen visually. The featureless disk or crescent shown in Fig. 5.5 is all you will normally observe of Venus through a telescope.

There are a couple of exceptions to the rule about the featurelessness of Venus. The first is that under exceptional observing conditions and with a good telescope, very vague, very low-contrast features can sometimes be seen.

---

[1] With its lengthy synodic period, Venus spends several months at a time a good distance to the east or west of the Sun. It is then a very bright star-like object to the naked eye and visible after sunset or before sunrise. It is thus sometimes called the evening star or morning star. Before it was realised that these two were the same object, they were given the names of Hesperos and Phosphoros by the ancient Greeks.

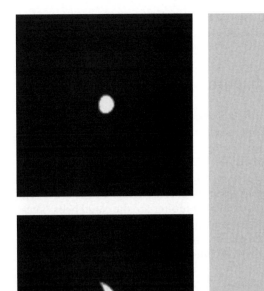

Figure 5.5 Phases of Venus – gibbous (*top*) and crescent (*middle*) phases – and showing the change in angular size (*bottom*) with differing distances from the Earth.

observed on the dark side of Venus, rather like the "Old Moon in the New Moon's arms" (see Fig. 4.3). It is controversial because it is very faint and ephemeral, and not all observers agree on its existence. If it is real, it could arise through charged particles from the Sun producing effects like the aurorae (Chapter 10), or perhaps it comes from lightning flashes lower in Venus' atmosphere.

Venus also undergoes transits,[2] though more rarely than Mercury. Venusian transits occur in pairs separated by 8 years around 7 June and 8 December. Each pair is then separated from the next by a 130-year interval. The next transits will be in 2004 (June) and 2012 (December).

5.2.3 Mars

Mars is an exciting planet to observe, although its synodic period of over two years means that opportunities to observe it are relatively infrequent. A further problem is that Mars' orbit is quite elliptical, so that at some oppositions its distance from the Earth can be over 100 million kilometres, while at others it may be only 55 million kilometres away. Its angular size at opposition can therefore vary from 14″ to 25″. The excitement arises through Mars' similarity to the Earth, and its thin atmosphere which allows its changing surface features to be seen. Those features, however, can be difficult to see and may be misinterpreted, as with the Martian canals.

Mars is interesting also as the one place in the solar system other than the Earth where there is a realistic possibility that indigenous life may exist or have existed in the past. Speculations on this topic range from H.G. Wells' lurid space invaders to possible fossils in meteorites originating from Mars. They seem unlikely to be settled until a manned expedition can visit the planet.

Spacecraft images show that Mars has a wide range of surface features, such as impact craters and basins, huge volcanoes, a gigantic rift valley and dried-up river beds. From the Earth, however, much less detail is visible. On an average night a few darker patches against the red disk may be discernible (see Fig. 5.6, *overleaf*), along with the polar caps. On a good night, with a good telescope, the patches become rather more detailed (see Fig. 5.7, *overleaf*), but still do not reveal the features shown by spacecraft.

These may become clearer if a deep blue or violet filter is used. They result from breaks in the upper layer of clouds on Venus allowing slightly deeper cloud layers to be seen.

Two other phenomena may be observed when Venus is close to inferior conjunction. The first is a bright circular ring around the dark planet. It arises because, although we are looking at the dark side of Venus, its thick atmosphere refracts light from the Sun. The refracted light then shows up as a ring. The second phenomenon is more controversial: it is called the *ashen light*. This is a glow

---

[2] Captain James Cook's voyage around the world in *Endeavour* (1768–71) had as one of its primary objectives the observation from Tahiti of the 1769 transit of Venus in order to obtain a more precise value for the Astronomical Unit.

Figure 5.6 Mars as typically seen through a small telescope on an average night.

Figure 5.7 Mars on a good observing night.

With practice at observing, the Martian features will become clearer and easier to see. This is because you will become adept at studying the images only in the brief intervals when the Earth's atmosphere clears and stabilises. In these intervals the theoretical resolution of the telescope (Eq. 2.5) may be reached, and you ignore the blurred images visible for the remainder of the time. It will then be possible to follow the changes in the polar ice caps with the Martian seasons and to see features in the atmosphere, such as dust storms. The rotation of Mars should be detectable, though if you observe at the same time every night, it will take some time before the rotation brings different features into view because Mars' rotation period is only 37 minutes longer than our own day.

Mars has two small satellites – Phobos and Deimos. At opposition their magnitudes are $+11.5^m$ and $+12.6^m$ respectively. Theoretically they should therefore be visible in 75 mm (3-inch) and 150 mm (6-inch) telescopes. However their angular distances from Mars at an average opposition are only 25″ and 60″, and the planet is some 400 000 times brighter than its satellites. So in practice the satellites become lost in the glare from Mars. A high-quality tele-scope of 300 mm to 400 mm (12 to 16 inches) aperture is usually the minimum needed to see Phobos and Deimos.

5.2.4 Jupiter

Jupiter is always a rewarding sight. At opposition it is 47″ across, but even near conjunction it is still 32″ across. Its synodic period is only a month longer than a year, so that it is regularly visible. We see only the top of its extremely thick atmosphere, but Jovian meteorology is very active, and the visible features are changing all the time. The major features visible on Jupiter are the cloud belts. These appear as darker and lighter bands stretching right the way round the planet, and can usually be seen even under poor observing conditions (see Fig. 5.8). Under better observing conditions, quite fine details in the belts become visible (see Fig. 5.9). The intensity and number of visible belts change on a time scale of a few months.

Jupiter rotates rapidly, but not as a solid body. The wind velocities of several hundred kilometres per hour present in its atmosphere are sufficient to give different

Figure 5.8 A typical view of Jupiter on an average night.

Figure 5.9 A view of Jupiter showing the sort of detail visible on a good night with a good telescope.

belts and regions different rotational periods. These therefore vary from 9 h 50 m to 9 h 55 m. Such a short rotational period means that, even in an hour or two, features will disappear around one limb and new features appear at the other limb. The centrifugal "force" arising from the rotation also makes the planet bulge at the equator, and even at low magnifications Jupiter can be seen to be distinctly elliptical.

In addition to the belts, spots of various types come and go on Jupiter. These are cyclone-type storms, but on a vastly greater scale than those that we get on the Earth. They often appear white, but other colours are possible. They can be several thousand kilometres across and last for several months. They are often moving through the atmosphere and so appear to have a different rotational period from their surroundings. One spot in particular, known as the Great Red Spot (GRS), seems to be semi-permanent. It is some 40 000 km long, and about 13 000 km in breadth, making it about the size and shape of the Earth's surface if that were to be peeled off, flattened and stuck on to Jupiter. The GRS changes its intensity in an irregular manner, first coming into prominence in 1878, but visible long before that.

Other features of various types are sometimes to be found on the planet. For many years a complex of spots and filaments, known as the South Tropical Disturbance, shared the same Jovian latitude as the GRS. The two features had different rotational periods and passed through each other several times with no apparent effect. The impact of comet Shoemaker–Levy 9 in 1994 produced a number of prominent spots which could be traced for many months after the events.

Jupiter has numerous satellites of which four are easily visible in small telescopes. These satellites were first seen by Galileo and are therefore known as the Galilean satellites; they are bright enough to be visible to the naked eye were they not lost in the glare from the planet. They can be seen as four bright stars strung out in a line along Jupiter's equatorial plane (see Fig. 5.10). Under the very best observing conditions, they can be seen as very tiny disks, but no features are detectable. They move rapidly since their orbital periods range from 1.77 days for Io, the innermost Galilean satellite, to 16.7 days for Callisto, the outermost one. Their mutual dance around the planet is fascinating to watch. They frequently transit across Jupiter or are eclipsed behind it. When in transit, their shadows are fairly easily visible on the surface of Jupiter. Very occasionally they may eclipse each other. Timetables of these various events are published in astronomy magazines, and astronomical society handbooks (Appendices 1 and 2).

Figure 5.10   The Galilean satellites (Jupiter is greatly over-exposed in this image in order to show the satellites).

The remaining dozen or so Jovian satellites range from 14^m to 20^m in brightness and are unlikely to be spotted using any telescope smaller than about 0.4 m (16-inch).

5.2.5 Saturn

Saturn is probably the single most beautiful sight in the sky, floating in the centre of its rings it is always breathtaking (see Fig. 5.11). Nothing else ever gives quite the same impression of the immensity of space. Like Jupiter, Saturn is a gas giant. On the planet therefore we see only

Figure 5.11   A view of Saturn as it may be seen on a reasonably good night.

the top of the atmosphere. Saturn has belts and occasional spots, but these are usually of fairly low contrast. Generally, Saturnian meteorology is far more tranquil than that of Jupiter.

The visible rings extend from close to the surface of Saturn to an outer radius of 140 000 km. Fainter rings are detectable in the infra-red or from spacecraft out to nearly 250 000 km. The thickness of the rings however is certainly no more than 10 km, and may be only a few metres. Thus every fifteen years or so, when the plane of the rings points towards the Earth, we see the rings edge-on, and they briefly disappear. These moments are opportunities to search near the planet for faint undiscovered satellites. At all other times we see the rings at greater or lesser inclinations. At maximum we may see them at an angle of 27° to the line of sight. The rings have several components, and rings C, B and A (moving out from the planet) should be visible, though ring C is rather faint. There is a dark division 4500 km wide between rings A and B, called Cassini's Division, which can be seen using a 75 mm (3-inch) or larger telescope except when the rings are nearly edge-on to us. Other divisions, such as Encke's division of ring A, may be detectable, especially around the times of maximum inclination of the rings. These minor divisions, however, are difficult to discern even under the best observing conditions.

Of Saturn's eighteen satellites, Titan is easily seen. Mimas, Enceladus, Tethys, Dione, Rhea and Iapetus should be visible with a 0.2 m (8-inch) or larger telescope. The remainder are very faint. Saturn's satellites are much more difficult to distinguish from field stars than the Jovian satellites because, except when the rings are near to being edge-on, the satellites are not strung out in a line. You therefore need to observe for some time in order to detect the satellites' motion, or look them up in an ephemeris, in order to be certain which are satellites and which are stars. Saturn's inclination also means that eclipses and transits of the satellites only occur when the Earth is near the ring plane.

5.3 Uranus, Neptune, Pluto and the Asteroids

Though Uranus and Neptune can be resolved, they are only a few seconds of arc across, and rarely can any details

be seen on them. Spacecraft images show that there is relatively little to be seen anyway. Of the satellites, only Triton, Neptune's largest satellite, is likely to be found with a telescope under 0.4 m (16-inch) in diameter. Pluto and the asteroids always appear star-like in a terrestrially based telescope.

Observational opportunities are therefore more limited with these objects in comparison with the other planets. The first task with any of them will be to find them! Apart perhaps from Uranus, they are likely to be too faint to see in the finder telescope. You will therefore have to find them in the main telescope. With accurate setting circles this will not be too difficult. Without setting circles, you will need to plot their positions on a star map and then star hop. Even when the telescope is pointing at the object, you will probably need a star chart to decide which points of light are stars and which is the object for which you are searching. This can even be true for Uranus and Neptune, because at the low magnifications used for star hopping, they may not be visible as disks.

Having found the object, its path around the heavens may be followed over a period of time. For asteroids (see Fig. 5.12), the motion may be detectable in an hour; for Uranus, Neptune and Pluto, observations separated by several days or weeks are likely to be needed to show the

Figure 5.12 An asteroid trail and the Sombrero galaxy (M104).

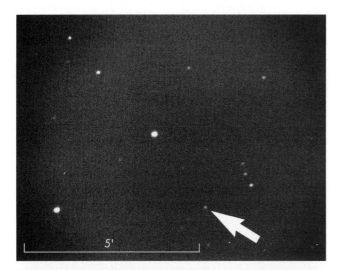

Figure 5.13 Pluto, showing its motion in 48 hours.

planet's movement (see Fig. 5.13). Near opposition, the rate of motion across the sky can be used to determine the distance of the object from the Sun using the graph shown in Fig. 5.14.

If you are able to add a camera or photometer to your telescope (Chapter 11), then the rotation of the asteroids may be studied from their light curves. Any irregularity shows up as a repeating pattern in the light curve at intervals separated by the rotation period. Adding filters to a photometer will then enable studies of the compositions of asteroids to be made by comparison with the common types of meteorite.

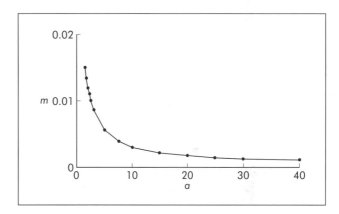

Figure 5.14 The rate of motion (*m*) near opposition of an object across the sky in seconds of arc per second of time as a function of its distance from the Sun (*a*) in astronomical units.

Chapter 6

Comets

6.1 Introduction

Until quite recently an appearance of a comet aroused feelings of dread, and was thought to presage calamities. While we no longer fear them, there is no doubt that a comet bright enough to be visible to the naked eye is a marvellous sight, with its shining coma and striking tail. Our word *comet* comes from "coma" meaning hair, and indeed the tail does look like hair blowing in the wind, (see Fig. 6.1).

The pursuit of new comets holds a particular fascination for some observers and part of the reason for this is their unpredictability. They are unpredictable both in terms of their times of appearance and in terms of how they "perform" – how bright they become and whether they develop conspicuous tails. Although comets have often assumed considerable significance for humans, on the cosmic scale they are modest objects, just a few tens of kilometres across. They attain their glory solely from the heating effects of the Sun, and from their eccentric orbits which result in short periods of intense solar heating. Let

Figure 6.1 Comet Hyakutake 1996B2, photographed on 26 March 1996. This was a striking naked-eye comet for a few nights, with a tail estimated to be up to 90° long by some observers.

us first consider the orbits and anatomy of comets, which govern their appearance, and then discuss their origins.

6.2 Cometary Orbits

6.2.1 Long-period Comets

Unlike the planets, most comets observed from Earth travel in highly elliptical orbits reaching out to thousands of Astronomical Units (AU), which means that they spend most of the time at great distances from the Sun. After a pass round the Sun at perihelion the comet is slowed down by the gravitational pull of the Sun as it recedes, so that by the time it reaches beyond the orbits of the planets it is travelling very slowly indeed. After perhaps 500 000 years or more it finally succumbs to the tiny but persistent pull from the Sun and passes aphelion to begin the equally long return journey. By the time it reaches the region of the terrestrial planets it has been accelerated to a speed that makes it dash around the Sun in just a few months

(see Fig. 6.2). It can be seen that such a comet spends most of its time in the very cold surroundings of the outer solar system, but is also subject to heating by the Sun for brief spells during its infrequent dashes through the inner solar system. Since the orbital periods are so long, these comets have not been recorded before and seem to appear at random.

6.2.2 Short-period Comets

There is also a family of comets with much smaller orbits having correspondingly shorter periods. These come close to the Sun much more regularly, can be seen from Earth at shorter intervals and are predictable. The orbit is usually still fairly elongated but in some cases can be almost circular. Because these comets thread constantly between the planets they are subject to frequent small changes in their orbits. Sometimes an encounter with one of the giant planets can cause a major orbital change. The term *periodic* is often used instead of short-period to describe these comets.

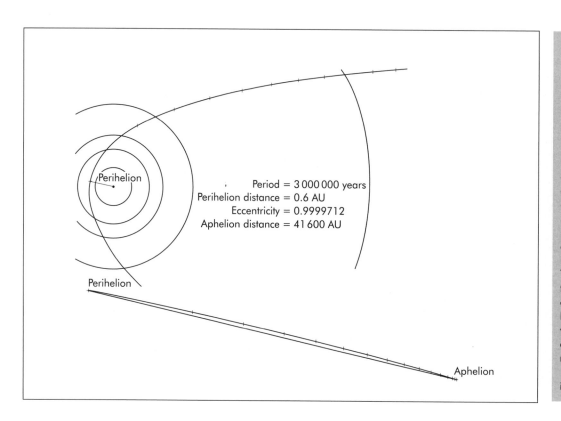

Figure 6.2 The orbit of a long-period comet. In the top diagram the comet is shown near perihelion: the marks on the path mark off intervals of 50 days. The orbits of the terrestrial planets and of Jupiter are shown. In the lower view the whole of the comet's orbit is shown. The marks are now at 100 000 year intervals.

6.2.3 Orbital Inclinations

Unlike the planets, whose orbits lie in planes that make only small angles with each other, a comet's orbit can be tilted at any angle. This means that they can be seen anywhere over the sky and are not confined to the Zodiac. The comet in Fig. 6.1, for example, was close to the Celestial North Pole when the photograph was taken. The direction of revolution may be in the opposite direction to that of the planets: this retrograde motion is described technically by an inclination of greater then 90°.

6.3 The Structure of Comets

6.3.1 Composition

Comets consist of a solid conglomeration of ices, dust particles and entrapped gas molecules. Spectroscopic evidence suggests strongly that most of the ice is ordinary water ice, with a significant amount of CO or CO_2 entrapped. Experiments that have collected interplanetary dust (presumed to have at least partly originated in comets) show oxides of silicon, iron and magnesium (that is, silicates). The solid body of the comet, or nucleus, is rather small, being a just a few tens of kilometres across.

6.3.2 Coma and Tail

When far from the heating effect of the Sun, comets are quiescent bodies with a low mean temperature. Any comet that approaches the Sun to within a few AU, however, experiences a strong heating of its outer surface. This heat cannot be conducted into its interior rapidly enough to prevent the surface temperature rising significantly. Then the volatile ices start to sublime (evaporate) off the surface, carrying off both embedded dust particles and small grains of ice. Sunlight reflected off these grains gives a bright appearance to the vicinity of the comet, forming the coma (see Fig. 6.3). The brightness drops off as the material expands and thins out and as the ice grains evaporate, leaving just the darker dust particles to reflect sunlight. At first the material expands in all directions, but soon the solar wind exerts a sufficient pressure on the particles that they are repulsed from the direction of the Sun and retreat from it. This gives a sharp-edged appearance to the coma in the direction of the Sun while in the opposite direction the material streams out to form the tail or tails. The visibility of the tail depends on the strength of sunlight and amount of material, both of which are greater when the comet is close to the Sun. Dust particles are sufficiently large that the solar wind does not affect them as much as gas and it is mainly radiation pressure that repels them from the Sun. This difference is reflected in the geometry of the tails – the high-speed gas tail is straight, whereas the slower speed of the particles in the dust tail leads to a curvature due to orbital effects. There

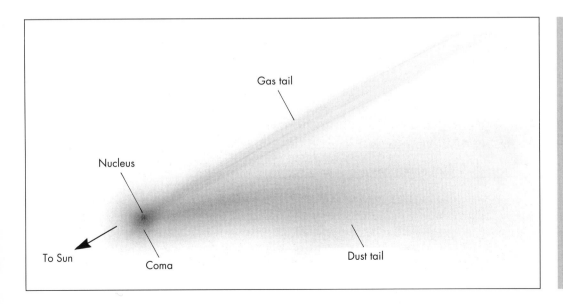

Figure 6.3 The main components of a comet. The relative size of the nucleus has been greatly exaggerated.

are also different mechanisms by which the tails are made visible: the dust simply reflects sunlight but the gas absorbs short solar wavelengths and re-emits visible light by the process of fluorescence. This causes a difference in the quality of the light, the dust having a similar colour to sunlight, whereas the gas tail's appearance is influenced by the main emission lines, which overall give it a bluish tinge. As the material flows away it interacts with the solar wind and local magnetic field to create a variety of effects. Sometimes twists appear in the ion tail, or disconnection events occur where a tail is "shed" and a new one appears.

6.3.3 The View from Earth

Given a comet of a particular tail length (possibly millions of kilometres) and brightness, the actual appearance in the sky is governed by the viewing geometry. A comet at considerable distance from the Sun will inevitably have any tail almost straight behind it as seen from the Earth, so that the angular size of the tail will be rather small. Similarly, a foreshortening of the tail will occur for a comet on the far side of the Sun and furthermore this position will, of course, confine observations to the bright twilight periods. The latter will also be true, in general, for those that approach the Sun most closely and thus exhibit the best activity. The most favourable viewing is when a comet passes the Earth relatively close by and at a greater solar distance than the Earth, so that it is near opposition to the Sun. There will be a further advantage if this occurs when the comet has passed perihelion and is receding from the Sun, as this is often the period when its activity is greatest. Occasionally an apparently sunward-pointing tail may be seen as a result of perspective. This usually takes the form of a narrow spike.

6.4 Origins

The question of the origin of comets is still unresolved, but it is believed that they formed in the region beyond the outer gas giants. The detection of objects in this *Kuiper belt* has been achieved recently, the most distant so far being at about 50 AU. Over the lifetime of the solar system, most of these small bodies have been perturbed by the planets into much larger orbits up to 20 000 AU across, forming the so-called inner comet cloud. They are then in a region where the attractive force of the Sun's

gravity is hardly greater than that of the Galaxy. They are now liable to be pulled about by the changing tides of the Galactic gravitational field until they are in orbits of typical size 80 000 AU. The comets in this region are collectively known as the *Oort cloud*. Encounters with passing stars cause a steady loss of these objects to interstellar space, while some perturbations result in the bodies plunging in towards the Sun so that they become the long-period comets.

Short-period comets are thought to arise in two ways: (1) a long-period object may have its orbit changed if it passes too close to a planet while in the inner solar system, leading to its "capture" into a smaller orbit; and (2) comets in the inner cloud or Kuiper belt can still be perturbed by the planets, and this sometimes again results in capture into a short-period orbit.

6.5 Famous Comets

A few comets are household names and a discussion of why this should be is of some interest.

6.5.1 Halley's Comet

The most famous of all, Halley's Comet, has an orbit carrying it from just 0.6 AU from the Sun at perihelion out to 18 AU at aphelion in a period of 76 years. Halley is the prime example of a short-period comet. Most times that the comet passes aphelion it also attains naked-eye visibility from the Earth; it has been seen at every return since AD66 for example. The fame of this comet (and also its name) arises because Edmund Halley noted that three comets seen in 1531, 1607 and 1682 had very similar orbits. He postulated that they were actually one and the same object and predicted its next return. The conditions for viewing naturally vary depending on the position of the Earth at each return, and the most recent apparition in 1986 was not very favourable. It was, nevertheless, a memorable return for a variety of reasons: (1) the comet was recovered over three years before perihelion by combination of an early CCD detector, computer-generated ephemeris and a large telescope (5 m); (2) spacecraft, including *Giotto*, sent back live images of the nucleus; (3) a world-wide collaboration between amateurs and professionals kept the most detailed watch of the comet yet seen; and (4) it is likely that the comet will continue to be observed all the way to aphelion in 2023.

Other frequently observed periodic comets are Encke (period 3.3 year), Schwassmann–Wachmann I which, unusually, has a nearly circular orbit of period 16 years but is worth watching because it often exhibits outbursts in brightness, and Temple–Tuttle (33 years), the source of the Leonid meteor storms.

6.5.2 Kohoutek's Comet

At the opposite extreme to Halley is comet Kohoutek, famous for having hardly been observed at all! It did exhibit activity and was observed by astronauts in the *Skylab* spacecraft but did not become a spectacle for naked-eye viewers as hoped. It should be noted, though, that it was still extensively observed by professional astronomers and very valuable results obtained.

In general, it is hoped that a comet that is brighter than average when still at a considerable distance from the Sun will become very conspicuous once it is subject to intense solar heating. Such an initial brightness, due simply to sunlight reflected off the surface, should imply a larger body that has a greater surface area. Unfortunately the process of sublimation which gives the comet its coma and tails can be erratic. The *Giotto* spacecraft took close-up pictures of Halley's comet in 1986 showing that its surface was rather dark. This implies that there can be a build-up of less volatile substances over the surface, which may shield the ices beneath from the Sun's heat. In this circumstance it would be expected that the comet would be less spectacular as less evaporation would take place.

Sometimes, however, enough heat will get through to cause a build-up of pressure sufficient to blast off part of the surface. This exposes fresh clean ice from which new material can evaporate in a jet. Alternatively, sporadic collisions between the comet and interplanetary debris may dig out a crater into fresh ice. For most comets there is no knowing whether they have large expanses of clean surfaces and can thus be particularly active, or if most of their surfaces are inactive.

6.5.3 Comet Shoemaker–Levy 9

Apart from sporadic landings of meteorites on Earth, comet Shoemaker–Levy 9 is the only observed example of a collision between solar system bodies. Although the collision was very unequal as the frail and small comet (already split up into many fragments by a prior close pass by

Jupiter) met the giant planet, the results were plainly visible with small telescopes. Jupiter sported black spots for some time afterwards. This example shows the importance of continued searches for new comets, as the advance prediction of the impacts enabled extensive observations by professional and amateurs alike. If the comet had slipped through the search net, invaluable information about Jupiter's atmosphere would have been lost. Furthermore, the eyesight of the first people to turn their telescopes on Jupiter after the event and report a sudden outbreak of black spots would have been called into question!

6.6 Nomenclature of Comets

With very few exceptions, such as some of those above, the naming of comets is governed by a strict system of nomenclature. The simplest designation is by year of discovery with a letter to show the order. Thus the sixth comet to be discovered in 1973, which was (the) comet Kohoutek had the designation 1973f. Once the orbit has been determined then the comet is given a second designation based on the year and order of passing perihelion, this time denoted by a Roman numeral. So 1973f is also 1973XII, being the twelfth comet to pass the Sun during 1973. Finally, and from the point of view of the discoverer most importantly, the name of the first observer or observers is given to the comet. As several observers have discovered more than one comet, some care may be needed in sorting out which object is referred to. For example 1973e was also called comet Kohoutek; comets with the same observer(s) name are sometimes distinguished by a final number as in Shoemaker–Levy 9. As an additional piece of information, the name of a comet that has been established as being of short period is prefaced by P/ for periodic. Thus the most recent designation for Halley's comet is P/Halley 1986III.

6.7 Observing Comets

6.7.1 Observing Information

Tables giving predictions for the positions of periodic comets can be found in publications from National Astronomical Societies (Appendix 1), such as the *Handbook of the British Astronomical Society*, together with the expected brightness. Details are also to be found in the popular astronomy magazines (Appendix 2) for the best objects or for recently discovered long-period comets that may become bright. If a comet is expected to be really bright then even the daily newspapers can be expected to carry details of where to find it in the sky. Another good source of information is the Internet, if you have access to a connection (Appendix 7).

6.7.2 What You Can Observe

Any chance of observing a comet should be seized upon, as you can never be sure what is going to happen. Every year several comets become bright enough to be easily observable in amateur-sized instruments. As each one is unique it is impossible to give a precise description of what to expect in a particular size of telescope. Most comet ephemerides will, however, carry a predicted total magnitude and as this amount of brightness is spread out over the whole object, it is likely that it will need to be a couple of magnitudes brighter than the limiting stellar magnitude of your equipment in order to be detected. Outbursts producing surges in brightness are not uncommon, however, so it may be worth having a look even if you do not expect the comet to be visible in your telescope. To pinpoint it use star hopping on a prepared chart, taking care to check that the equinoxes of the chart and comet ephemeris are the same (both 2000.0, for example). If they are not, you will have to apply a precession correction to the comet's position. A comet that is passing relatively close by the Earth will move across the sky quite rapidly, so it will then be necessary to plot the comet's place at the time you expect to observe it. Inspection of the daily positions given will show if this is required, and if so you will have to interpolate between positions. If the brightness is predicted to be greater than about 5th or 6th magnitude, then elaborate preparation will be unnecessary as the comet should be readily picked up by sweeping with the finder or binoculars.

The most obvious feature will normally be the coma. Examine it carefully. Does it look symmetrical or does it have a sharper boundary on one side? If so, can you discern any extension of the less-sharp side into a tail? Placing the brightest part of the coma to one side out of the field of view may help here. Estimate the visible angular size of the coma by comparing it with the eyepiece field of view. Now observe more closely. Is the

cometary body itself visible as a bright nucleus inside the coma? How sharp does it appear – is it star-like and clearly seen, or faint and a bit fuzzy? Variations in the appearance of the nucleus can arise because of its rotation and/or because of variations in the rate of activity. If the nucleus is of irregular shape (such as Halley's was shown to be by *Giotto*), rotation will make larger or smaller areas visible in turn and so vary the apparent brightness. When the activity increases there is an increased density of particles in the coma which impedes the view through to the nucleus. Hence the nucleus can be clear one night and much less apparent the next. Finally, can any features be detected close to the nucleus, such as jets from active points on the surface? Again these will vary from day to day. An accurate sketch of any features or a photograph or CCD image would be valuable. Examination of a series of these can lead to a determination of the rotation period of the comet if most of its activity is confined to a few isolated jets. Except for very short exposures it will be necessary to guide on the comet to compensate for its drift against the background stars. This normally means that a separate guiding telescope must be used to guide on the nucleus. CCDs allow the build-up of a long exposure by stacking together shorter exposures, so this may be another alternative (see Fig. 6.4).

Figure 6.4 An image of comet Mueller 1993a produced by stacking several exposures. Note the star-like nucleus. The trails of the background stars, due to the comet's rapid motion across the sky, are a characteristic feature of all comet pictures of substantial exposure time.

An image with background stars and cometary nucleus both visible can be used to determine the position of the body on the sky and thus help in determining the parameters of its orbit.

Look further afield along the track of the tail to see if you can discern any features. The finder or a pair of binoculars may be very useful here. They will help to indicate the direction in which the tail or tails lie, and the lower magnification will usually enhance their visibility.

A more difficult problem is to determine the brightness of the comet, as its appearance is so different from that of stars acting as comparisons. The method is to de-focus stars of known magnitude until one forms a blob of similar brightness and size to the comet. The overall magnitude of the comet must then approximate to that of the star.

6.7.3 Discovering Comets

The discovery of comets requires a particular dedication that only a few people will wish to display. It requires on average hundreds of hours of searching before success is achieved. Many comets that approach in a direction well away from the Sun are picked up during routine observations by professional astronomers, so the best bet is to go for those that happen to arrive from behind the Sun. They will emerge from perihelion into the morning or evening twilight sky, so that is the place to search. It has to be stressed, however, that extreme caution is required before breaking out the champagne if you think you have succeeded (see also Section 2.12). That fuzzy object may be a distant star cluster or gaseous nebula, or an even more distant galaxy or indeed a comet that is already known about – but there again it might be *your* comet!

Chapter 7

Stars

7.1 Introduction

"When you have seen one star you have seen the lot" – this is a commonly encountered sentiment among many amateur astronomers who prefer to spend their time with the apparently more rewarding extended objects such as planets, nebulae and galaxies. There is also a certain amount of truth in it, for no telescope will show a star as anything other than the point source that can be seen by the naked eye.[1] However, with a bit of effort, stars can be as interesting to observe as the most spectacular nebulae, and furthermore they provide one of the few areas where an observer with a small telescope can contribute to genuine astronomical research. In the rest of this chapter we hope therefore to convince you of the opportunities for the stellar observer.

7.2 Brightness

7.2.1 Magnitudes

With stars, which are point sources,[2] the telescope does increase their brightnesses (unlike the situation for extended sources – Section 2.5.1). The increase in brightness is given by the light grasp (Eq. (2.2)). For historical reasons, astronomers use a rather complex scale for measuring the brightnesses of stars and other astronomical objects – called the magnitude of the object. The magnitude is related to the brightness (that is, the amount of light energy received from the star per unit area) by Pogson's equation;

$$m_1 - m_2 = -2.5 \log_{10} (B_1/B_2) \tag{7.1}$$

where m_1 and m_2 are the magnitudes of stars 1 and 2,
B_1 and B_2 are the brightnesses of stars 1 and 2.

You should not be put off by this equation, because it was defined only to give a precise formulation to the practice of making eye-estimates of the brightnesses of stars. The magnitude scale is therefore ideally suited to and based on visual observations. The equation, however, does reveal several peculiarities of the magnitude scale. Firstly, it is not an absolute scale; the magnitude of one star is obtained by comparison with that of another. So at least one star must arbitrarily be assigned a magnitude before that of any other star can be determined. In practice, the scale is chosen so that stars of magnitude 6 are those just

---

[1] Using special techniques on large telescopes it is possible to resolve some of the larger and nearer stars, but this is a long and complex process. Even on the largest telescopes, or with the Hubble space telescope, it remains impossible to resolve stellar disks directly.

[2] The definition of a *point source* in this context is that the image is smaller than the detecting element. For visual work, the detecting elements are the rod and cone cells in the retina of the eye, some few tens of microns across, giving a resolution of 3 to 10 minutes of arc. For reasonably average atmospheric conditions, with stellar images 2 seconds of arc across, stars therefore cease to be point sources when the magnification exceeds somewhere between ×100 and ×300, depending on individual physiologies.

visible to the naked eye from a good observing site. Various stars, known as Standard Stars, are identified around the sky with precisely known magnitudes to enable those of other stars to be measured. For visual work, any non-variable star whose magnitude can be obtained from a star catalogue or atlas (Appendix 2) will normally be sufficient to act as a standard for comparison.

The second peculiarity of the magnitude scale is shown by the presence of the negative sign on the right-hand side of Eq. (7.1). This results in the value of the magnitude being SMALLER the BRIGHTER the star. Indeed the brightest stars and other objects have negative magnitudes. Thus the Pole Star (Polaris, α UMi) which has a magnitude of 2.0 is fainter than Dubhe (α UMa) whose magnitude is 1.8. Sirius (α CMa), the brightest star in the night sky, has a magnitude of –1.5.

The final peculiarity of the scale is shown by the presence of the logarithm on the right-hand side. A difference of one magnitude between one star and another thus results not in a constant **difference** in their energies, but in a constant **factor** between their energies. That factor (this number is actually $10^{0.4}$) is ×2.512. If two stars differ in brightness by one magnitude, the brighter is therefore 2.512 times more luminous than the fainter whether the two stars are of magnitudes 1 and 2, or 20 and 21. If two stars differ in brightness by two magnitudes, then their energies differ by a factor of $(2.512)^2$, and if they differ by three magnitudes, then their energies differ by a factor $(2.512)^3$ etc. Table 7.1 lists the relationship in more detail. Star catalogues and atlases (Appendix 2) list the magnitudes of stars (and other objects such as nebulae and galaxies); a few examples are given in Table 7.2 to provide a "feel" for the way the system works.

The magnitudes listed in Table 7.2 are *apparent visual magnitudes*; that is, they are the magnitudes of the objects as they appear in the sky to the eye. In star catalogues they will normally be denoted by "m", "m_v", m_{pv} (for photovisual), or sometimes by "V". If stellar brightnesses are measured using detectors such as CCDs, photographic emulsion, photomultipliers etc., rather than through eye estimates, the resulting magnitudes will normally have different values. This is because the eye is most sensitive to the yellow-green part of the spectrum, but other detectors may be looking at the stars in the ultra-violet, blue red or infra-red parts of the spectrum. Star catalogues will normally label such magnitudes as "m_p", "U", "B", "R", "I" (these symbols stand for photographic, ultra-violet, blue, red and infra-red magnitudes respectively) and these values should NOT be used as the magnitudes of stars if you are working visually.

Table 7.1. The relationship between stellar magnitudes and energy

Magnitude difference $(m_1 - m_2)$	Energy ratio (B_1/B_2)
0.1	1.1
0.2	1.2
0.3	1.3
0.4	1.4
0.5	1.6
0.6	1.7
0.7	1.9
0.8	2.1
0.9	2.3
1.0	2.5
2.0	6.3
3.0	15.9
4.0	39.8
5.0	100 (exact)
10.0	10 000 (exact)
15.0	1 000 000 (exact)
20.0	100 000 000 (exact)
50.0	100 000 000 000 000 000 000 (exact)

Table 7.2. Examples of stellar and other magnitudes

Object	Magnitude
Sun	–26.7
Full Moon	–12.7
Venus (maximum)	–4.3
Jupiter (maximum)	–2.6
Mars (maximum)	–2.02
Sirius A	–1.45
Betelgeuse	–0.73 (variable)
Polaris	+ 2.0
Uranus (maximum)	+ 5.5
Faintest object visible to the unaided eye	+ 6.0
Faintest object visible in a 0.3 m telescope[a]	+ 13
Faintest object visible in a 1 m telescope	+ 16
Faintest object visible in a 10 m telescope	+ 21
Faintest object detectable using the very best of modern techniques	+ 28

[a] The limiting visual magnitude through a telescope is given by Eq. (2.3).

Star catalogues will also normally list another type of magnitude: the *absolute magnitude*. This will be denoted by an upper-case letter such as "M", "M_v", "M_p", etc. The absolute magnitude is defined as "the magnitude of the object if it were at a distance of 10 parsecs". It is used because it is directly related to the actual luminosity of the object. The apparent magnitude depends on both the actual luminosity and the distance of the object. Thus Sirius A (α CMa) and Mintaka (δ Ori) have apparent magnitudes of –1.5 and +2.2 respectively, making Sirius 3.7 magnitudes (×30) brighter than Mintaka. However Mintaka is nearly 200 times further away from Earth than Sirius. Their respective absolute magnitudes are thus +1.4 and –6.1, making Mintaka actually 7.5 magnitudes (×1000) brighter than Sirius. Useful though the absolute magnitudes are for many purposes, it is the apparent magnitudes that are needed by an observer. Further details of absolute magnitudes are therefore left to the interested reader to pursue in sources listed in Appendix 2.

7.2.2 Estimating Visual Magnitudes

Equation (7.1) has probably given the impression that the last thing you would ever be likely to get involved with is determining magnitudes! However, although the equation is needed when a CCD or other similar detector is used to measure the brightnesses of stars, it can be forgotten about completely for eye estimations. This is because the eye responds to differences in brightness in a logarithmic fashion. (The logarithmic response of the eye is of course the reason why the magnitude scale is logarithmic. A similar effect occurs for differences in sound level and has resulted in the logarithmic decibel scale.) Thus if an unknown star appears to the eye to be halfway in brightness between two stars of magnitudes 3 and 4, then its magnitude is automatically halfway between them at 3.5^m, if it appears to be three-quarters of the way towards the brighter star, then its magnitude is 3.25^m, and so on.

Visual magnitude estimations are therefore quite easy to make, and with some practice can be made reliably to $\pm0.1^m$. To estimate the magnitude of an unknown star, at least one star of known magnitude must be visible in the same field of view. The known star should be of not too different a brightness from the unknown one. It would be very imprecise, for example, to use Sirius A (-1.5^m) to try and estimate the magnitude of Sirius B ($+8.7^m$). When using just a single comparison star, you will first have needed to practise using pairs of stars of known magnitude differences, in order to estimate the difference between your known and unknown stars. Once you have trained yourself however, it is straightforward to estimate that the unknown star is, say, half a magnitude brighter than the known one.

Using a single comparison star requires experience before consistent and reliable results can be obtained. Where possible, therefore, it is much better to have two comparison stars, one of which is slightly brighter and the other of which is slightly fainter than the unknown star. Even without much practice it is then possible to get good magnitude estimates by judging whereabouts between the brightnesses of the two known stars, the unknown one lies (see above).

Where possible, the comparison star(s) should be similar in temperature[3] to the unknown one. This is because the temperature of a star affects its colour; cool stars are reddish, medium-temperature stars, white, and very hot ones, bluish. Since the eye is most sensitive to the yellow-green part of the spectrum, inaccuracies in estimates of magnitudes will result when stars of different temperature are compared.

Another problem area which can affect eye estimates of magnitudes is vignetting in the telescope or eyepiece. Vignetting occurs when parts of the structure of the instrument obstruct some areas of the field of view, or when not all parts of the optics are fully illuminated. It is most likely to occur with the more elaborate designs of telescope and eyepiece and/or when stops are used in the system. Vignetting will usually affect the outer parts of the field of view most and will result in objects appearing to be fainter when at the edge of the field of view than when they are centred. Unless you are convinced that no vignetting is occurring in your telescope and eyepiece, you should adjust the position of the telescope until the known and unknown objects are equal distances from the centre of the field of view, before estimating their differences in brightness.

It is often difficult to find suitable comparison stars for the unknown star. Various devices using tilting mirrors etc. can therefore be added to the telescope to enable two widely separated stars to be brought into the same field of view. It also possible to use adjustable filters or other devices to equalise the brightness of the stars. Such refinements are discussed for example in *Astrophysical Techniques* (Appendix 2, Section A2.5), but are well beyond the scope of this book.

---

[3] The star's temperature is obtained from its *spectral type* or *colour index*. These quantities will be listed in star catalogues along with other data on the star. The precise relationship of these quantities to stellar temperature is, however, beyond the scope of this book, and the interested reader is referred to sources in Appendix 2.

7.3 Variable Stars

7.3.1 Observing Variable Stars

There is little point, except when practising, in estimating the magnitudes of stars chosen at random. Most of the stars visible in a small telescope will already have known magnitudes. The interest in magnitude determinations occurs, therefore, for those stars which vary in brightness, usually known as photometric variable stars (see Fig. 7.1).

It is in the area of variable-star observation that the amateur can compete with the professional astronomer and make a real contribution to astronomical research. Variable stars by their very nature call attention to themselves and are likely to be among the more interesting stars in the sky. Additionally, the nature of the variation can sometimes reveal far more information about the stars than it is possible to obtain for the sedate majority of non-variables. Variable stars form only a small proportion of the total number of stars, but nonetheless this still leaves nearly 24 000 stars listed in the *General Catalogue of Variable Stars*. The majority of these stars will not have been studied in any detail. Indeed, in some cases the only observations will be the few survey photographs which revealed the change in the star's brightness. With so many stars to cover, even those that have been studied will probably only have been observed over a single one- or two-year period, and this may have been many decades ago. There is therefore considerable scope even among the brighter variable stars for original and follow-up work to be undertaken.

You can simply choose your own programme of work if you decide to study variable stars, selecting one star, or perhaps a number of stars of similar type, to monitor over an extended period. The effort involved is not large, since with most variables it is sufficient to estimate their brightness every few nights or even just once a month. Once you are familiar with the variable star and its comparison star(s), estimating its magnitude is only a few minutes' work at most. If you select this route, then you will need to research your star(s) yourself both for their individual properties and for what may be understood from their variations. This will require access to a good astronomical library which has the major astronomical research journals. To obtain these, you can try approaching a nearby university or research institute, since such places will often allow genuine enquirers to use their libraries, or you may need to join your national astronomical society (Appendix 1).

An alternative, and perhaps preferable approach to going it alone, is to join with others in an established research programme. Your national or perhaps more local astronomical society (Appendix 1) will have a group of people already working on variable stars. Usually these observing programmes will be coordinated nationally and even internationally so that they complement each other without duplicating work. There are several advantages to joining an established programme. Firstly, the variables needing to be observed and suitable comparison stars for them will already be known. Secondly, it is much easier to get a continuous coverage of the light curve, and to get more precise measurements, if several people are observing a star, and finally, the background research on the stars and the astrophysics to be obtained from their study will already have been done.

7.3.2 Types of Variable Star

While lack of space precludes coverage of the details of the myriad types of variable star here, an outline of the major groupings and of their properties will be useful and perhaps enable the reader to select an area of possible interest for further study. Variable stars are divided into the *extrinsic variables*, where the change in brightness arises from some process external to the star, and the *intrinsic variables*, where the star itself is changing in some way.

The extrinsic variables are mostly eclipsing binary stars. These are two stars in orbit around their centre of mass, and with their orbital plane lying close to the line of sight. One star therefore passes in front of the other, eclipsing it and reducing the brightness of the combined system, twice per orbital period. The light curve shows a

Figure 7.1 Two images superimposed side by side of the photometric variable, RZ Cas, showing it at maximum and minimum.

series of regular dips or minima. The eclipse of the hotter star produces the deepest reduction in brightness, and this is called the primary minimum. The eclipse of the cooler star produces a secondary minimum. The main interest in the study of eclipsing variable stars arises because they provide one of the few ways in which the masses of stars may be determined.

The intrinsic variables include a huge range of sub-types. Those which might be included in a possible observing programme could include:

1. *Cepheids*: where the star is pulsating in size with periods from 1 to 100 days, so changing its brightness, and where the light curve enables the distance of the star to be found.

2. *Long-period or Mira-type variables*: these have large changes in brightness on time scales from months to years.

3. *Dwarf and recurrent novae and flare stars*: stars which undergo periodic explosive increases in brightness on time scales from a few weeks to tens of years.

4. *R CrB stars*: these have sudden large decreases in brightness, possibly due to the formation of large quantities of dust in their outer layers.

5. *Irregular variables*: a large and heterogeneous group where the brightness changes show no evidence of periodicity.

The cataclysmic variables also belong to the intrinsic variable class, but are sufficiently important to be dealt with separately. The group includes the novae and supernovae. These are stars which undergo enormous explosions and which therefore change their brightnesses by huge amounts.[4] A supernova (see Fig. 7.2), which completely destroys the original star, can peak at a brightness several thousand million times that of the Sun. Novae and super-novae are among the most interesting and exciting objects in the sky, but their occurrence cannot be predicted. Most novae and supernovae (and also comets – Chapter 6) are therefore discovered by amateur astronomers specialising in searching the sky for them. Becoming a nova hunter is a lengthy process, because you first have to become thoroughly familiar with the appearance of the sky through a low-power telescope, and this process of familiarisation is likely to take several years. Thereafter, you need to survey

---

[4] The name "Nova" derives from "Nova Stellarum" or New Star, because prior to the invention of the telescope a bright nova or supernova would appear in the sky where no star had previously been seen.

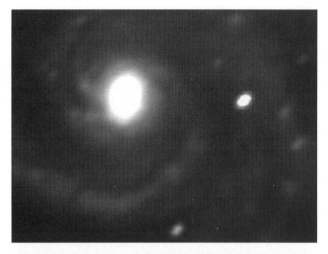

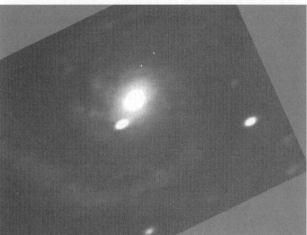

Figure 7.2 A supernova in M51 (compare also Fig. 9.9): (*top*) before the supernova; (*bottom*) supernova appears at the edge of the nucleus.

the sky at every opportunity, trying to find the first signs of a nova, supernova or comet. The rewards for success are commensurate with the effort involved – in the case of the discovery of a comet, your name being given to that comet in perpetuity. The procedures for reporting discoveries, and also cautions against false discoveries, are discussed in Section 2.12.

A final note on variables concerns the nomenclature used for them. This is complex and confusing with many exceptions. The basic principle, however, is to extend the Bayer system (Chapter 1). With that system, the Greek letters are used for the first 24 stars of a constellation, followed by lower-case and upper-case Roman letters, as in α Her, e Vel and B Cen. The highest letter that had been utilised in this way by the middle of the nineteenth century was Q, so Argelander used the letters R to Z to denote the brighter variable stars in each constellation, as in R Scu and V Boö. Many more than nine variable stars were soon found in all constellations however, and so double upper-case Roman letters were used. These however did not start at AA, but at RR, going from RR to RZ, as in RR Lyr and RZ Cas. This gave another nine labels. Thereafter the following sequences were used: SS to SZ, TT to TZ and so on up to ZZ, then AA to AZ (omitting J), BB to BZ, CC to CZ and so on up to the 334th variable in the constellation which was labelled QZ. With the discovery of the 335th variable in a constellation, reason prevailed and this was just designated V335 followed by the constellation, as in V484 Ori and V383 Lyr, rather than going to three letters. Some variables however had already been given names before Argelander's system came into use, and so β Lyr, β Per and P Cyg are actually variable stars. Novae and supernovae are also labelled differently. Novae are initially named by constellation and year as in Nova Serpentis 1970, later receiving a normal variable designation, so that Nova Ser 1970 is now known as FH Ser. Supernovae are labelled by year and a sequential upper-case Roman letter. So that SN 1987 A was the first supernova discovered in 1987. There are many other exceptions and special cases, but this is probably enough for now!

7.4 Visual Double and Binary Stars

Double stars are two stars which appear close together in the sky. Their angular separations can be anything from a few minutes of arc to less than milli arc-seconds, though at the latter end of the range only special techniques will reveal the duplicity. Double stars can occur because two stars happen to lie in nearly the same direction as seen from Earth, though they are at very different distances away from us. There is then no physical connection between the stars. They may also occur when the two stars are close together in space and are gravitationally bound to each other. These latter stars are usually called *binary stars* rather than double stars, and they will be orbiting around their common centre of mass. If the two stars of a binary system can be seen directly, then it is a visual binary. There are also many binary systems which appear

to be single stars when viewed directly, but where brightness or spectral variations show them to be binary. These latter cases are called photometric binaries (see eclipsing binaries above) and spectroscopic binaries, and are of no further concern here.

Triple, quadruple and higher-order combinations of stars can occur, with the larger the number of components, the greater the probability that the system is physically connected. The progression leads on to galactic and globular clusters (Section 7.5) and eventually to galaxies (Chapter 9), but here we are concerned with systems containing only a few stars. Some very complex combinations can occur, such as:

1. ε Lyr (also called the "Double-Double") which comprises two visual binaries with separations of 2.3″ and 2.6″, which in turn are separated from each other by 208″.

2. Mizar and Alcor (ζ and 80 UMa – Fig. 2.19) where the two named stars are separated by 11.5′ and can be seen by most people to be double with the naked eye. Mizar is then a visual double with a separation of 14.5″, and the brighter component of this double is a spectroscopic binary with a separation (if we could see it) of 0.04″.

3. Sirius (α CMa) which is a visual binary with a separation ranging from 3″ to 12″, the brighter component, however, some 10 000 times brighter than its companion.

Double stars are usually observed for three main reasons. The first is simply for their beauty. Complex systems such as ε Lyr, or systems where there are strong colour differences between the components (Table 7.3) are well worth observing in their own right. The second reason is as a test for your telescope and of your observing ability. Close double stars (that is, those with separations of 0.5″ to 5″) may be used to test the resolving power (Eq. (2.5)) of your telescope, and your ability to observe. Finally, you can also use them as a test for the quality of the atmospheric turbulence (seeing, scintillation and twinkling – Section 2.9), since on poor nights even widely separated doubles may not be resolved. A selection of double stars with differing separations will provide a quantitative measure of the observing conditions based on the closest separation that can be resolved.

Finally, for those few visual binaries whose periods are short enough to enable their orbits to be plotted out in less than a few decades, observations of the complete orbit leads to accurate determinations of the masses of the two stars.

Table 7.3. Colour contrasts in double stars

Star	Number of components	Colours
γ And	2	orange, blue
ν And	3	red, yellow-green, blue
ε Boö	2	yellow, blue
ξ Boö	2	red, deep red
α CVn	2	yellow, blue
ι Cas	3	yellow, red, yellow-green
β Cyg	2	yellow, blue
γ Del	2	yellow, yellow-green
α Her	2	red, yellow-green
β Ori	2	white, blue
η Per	2	red, blue
α Sco	2	red, yellow-green
β Sco	2	yellow-green, blue

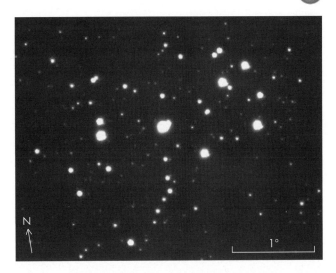

Figure 7.3 The appearance of the Pleiades galactic cluster (M45) on a good night.

We have seen in the previous section that multiple star systems can contain 2, 3, 4 or more components. At some point as the number of components increases, a multiple system will become called a star cluster, though that point is ill-defined. Certainly, however, 10 or 20 or more stars gravitationally bound together would be called a cluster. When the number of stars rises to 10^8 or more, clusters start to become galaxies (Chapter 9). This section is therefore concerned with aggregations of stars between those two limits. In that range there are two main structures to be found – *galactic* or *open clusters*, and *globular clusters*.

7.5.1 Galactic Clusters

The best known example of a galactic cluster is easily visible to the naked eye and it is the Pleiades (M45 – Figs 7.3, 1.17 and 1.18). This cluster is also known as the Seven Sisters, though in fact only six stars can normally be seen with the naked eye. Through a telescope three or four hundred stars become visible, and the cluster may contain as many as 3000 stars in total. The stars are linked together by gravity and orbit around the galaxy as a single unit. The cluster has recently formed from an H II region (Chapter 8), and will eventually break up into individual stars, binaries and multiple systems. Some of the gas remaining from the H II region is visible on long-exposure images of the Pleiades as a blue reflection nebulosity around the brighter stars. There are several variables worth observing in the cluster, including Pleione (28 Tau, BU Tau) which, were it to have been brighter in the past, could have been the seventh star of the seven sisters. The Wild Duck cluster (M11: Figs 7.4 and 7.5, *overleaf*) is an almost comparable fine sight for southern observers, containing perhaps up to 600 stars. There are many other galactic clusters, and some of the brighter ones are listed in Appendix 4, while others may be found from star charts and catalogues (Appendix 2).

Observing galactic clusters has no special requirements. For the angularly larger ones like the Pleiades, a very low-power, wide-angle eyepiece will be needed if most or all of the cluster is to be seen in one go. More distant clusters may need high powers and good observing conditions if they are to be resolved into stars, and the most distant ones will remain nebulous at all times. A spectroscope is needed for the latter to show that they are indeed composed of stars and are not a gaseous nebula. Some of the angularly very large clusters listed in Appendix 4 (Table 4.1), such as Ursa Major, do not appear directly as clusters. Their movements through space when obtained from their proper motions and radial velocities, however, show that the stars are all moving as one around the galaxy.

Other types of clusters are to be found, such as the T Associations. These do not, however, normally show as

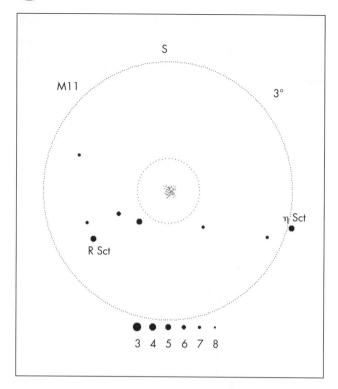

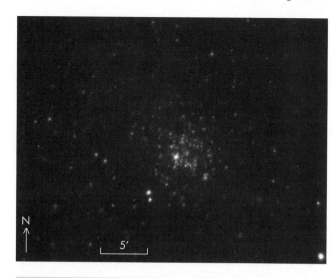

Figure 7.5 The appearance of the Wild Duck galactic cluster (M11) on a good night.

clusters when viewed directly. They are groups of stars of similar spectral types (T Tauri stars in the case of T associations), but without a noticeably higher space density than normal. They only become apparent when studied spectroscopically.

7.5.2 Globular Clusters

Globular clusters are small satellite galaxies found in large numbers around larger galaxies. They contain from a few tens of thousand to over a million stars. They are spherical and the stars are very strongly concentrated towards their centres. The Milky Way's globular clusters are distributed in an enormous spherical halo centred on the galaxy. Many clusters are therefore to be found in the southern hemisphere around the galactic centre in Sagittarius. The two brightest ones, ω Cen and 47 Tuc (C80 and C106: Figs 1.22 and 1.23) are clearly naked-eye objects. Several others

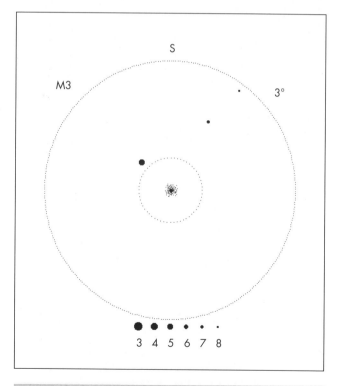

Figure 7.6 Finder chart for M3.

---

[5] The centre circle on these finder charts shows approximately the field of view of a low-power eyepiece on a small telescope.

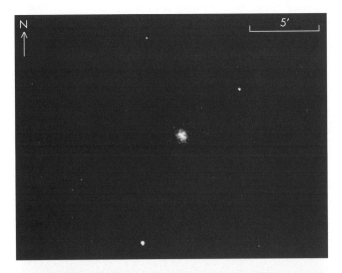

are marginally visible on good dark nights. Through even a small telescope, stars in the outer reaches of some of the nearer globular clusters (see Figs 7.6 and 7.8) can be resolved (see Fig. 7.7 and Fig. 7.9, *overleaf*). The main parts of most globular clusters, however, appear as circular nebulosities. Observing globular clusters is therefore analogous to observing gaseous nebulae (Chapter 8) and galaxies (Chapter 9). Some of the more prominent globular clusters are listed in Appendix 4.

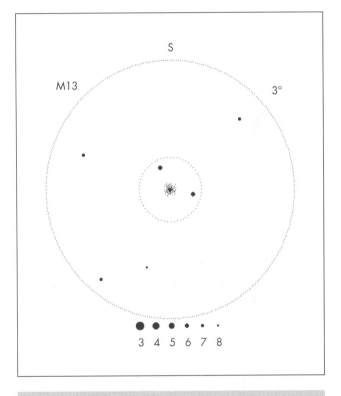

Figure 7.8   Finder chart for M13.

Figure 7.7   The appearance of M3 through 3-inch (75 mm), 6-inch (150 mm) and 12-inch (0.3 m) telescopes.

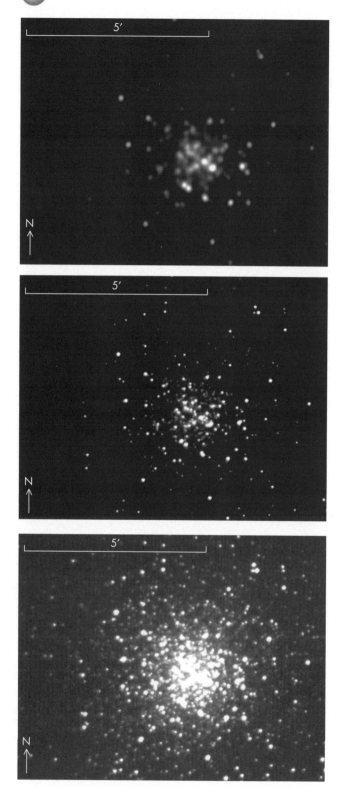

Figure 7.9 The appearance of M13 through 3-inch (75 mm), 6-inch (150 mm) and 12-inch (0.3 m) telescopes.

Chapter 8

Nebulae

8.1 Introduction

The study of nebulae has become one of the most dynamic areas of astrophysical research, as the ways in which they are involved with the lives of stars have become clearer. The use of radio telescopes and infra-red detectors has enabled the interiors of emission nebulae to be studied in detail, revealing a multitude of processes hitherto hidden from view. The word *nebula* is the Latin for mist, and related to the Greek for cloud, and this sums up the typical appearance of a nebula very well: a misty patch in the sky. Although the term originally encompassed galaxies and star clusters, it is now reserved for five types of object: dark nebulae, reflection nebulae, emission nebulae, supernova remnants and planetary nebulae.

8.2 Gas and Dust Clouds

Sprinkled quite thickly throughout the interstellar medium of the Galaxy are to be found volumes with denser than average concentrations of gas, forming the gas clouds. Spectroscopic analysis shows that this is mainly hydrogen, with about a quarter of it helium. Intermingled with the gas is a small amount of dust which acts as a very efficient absorber of light, preventing a good view of the interior of the cloud. Fortunately the dust does not impede the passage of infra-red or radio waves to the same degree, and observations at these longer wavelengths show the presence of many molecules and embedded warm objects. The complexity of some of these molecules shows that inside the clouds the temperature is rather low, as otherwise they would break apart. The presence of warmer objects, therefore, indicates localised sources of energy, and the only plausible mechanism for these is gravitational contraction of the densest parts of the clouds into protostars. To bolster this conclusion, infra-red views show that many nebulae have star clusters hidden inside them, and of course even optical observations have long shown that a young star cluster is often closely associated with a nebula. All the visual manifestations of emission, reflection and dark nebulae stem from these basic facts (see Fig. 8.1, *overleaf*).

8.3 Dark Nebulae

8.3.1 Introduction

Also known as absorption nebulae, these are seen as dark islands silhouetted against background stars, or against emission nebulae. Their appearance is simply due to their dust content, which absorbs light from objects beyond. In some cases the absorption is moderate and some stars can be seen, in others the sky appears completely black. This variability is to be expected, as the constant orbital motion of everything around the Galactic centre and the effects of star formation and evolution stir up the interstellar medium to produce a chaotic jumble in which the

93

Figure 8.1 Nebulae in Orion, photographed with a 200 mm Schmidt camera. The positions of several nebulae of the different types are indicated.

density of material varies from place to place. The amount of absorption depends on the density of dust in the nebula and its depth, that is on the total number of dust grains along the line of sight. The density is normally rather low and so the large dimming is a consequence of the huge total dimensions of many of these clouds. Some dark nebulae are readily apparent to the naked eye, for example the Cygnus rift in the northern Milky Way and the Coalsack in the far south. Others are more subtle,

either because they are only seen against a faint background, because they are small, or because their ragged boundaries do not stand out clearly to the eye.

8.3.2 Nebulae

The Horsehead Nebula   RA$_{2000}$: 05 h 40.9 m   Dec$_{2000}$: –02°28′ (see Fig. 8.2)

This is probably the most famous of the dark nebulae and exemplifies the tendency of astronomers to give descriptive names to objects of all sorts. It is also known more prosaically as Barnard 33. Illustrations of it can be found in many books. It is a dark arm of material seen against the background of the emission nebula IC 434, a streak running south from zeta Orionis. It is quite easy to photograph with fast optics, being visible as a notch in IC 434 even in pictures taken with a 50 mm lens. The clear shape, though, requires 300 mm focal length or more. Visually it is a difficult object, and the best chance of success is to use an H-beta line filter, and to ensure that your optics are clean to reduce scattered light from nearby stars. The nebula is of moderate size, 6′ × 4′, with the longer dimension east–west. A dark nebula of width similar to the Horsehead but twice as long is to be found a little to the north, running north–south across the emission region NGC 2024. This latter nebula lies immediately to the east of zeta Orionis, and would be an easy object if it were further away from the star. (Note the brightest star

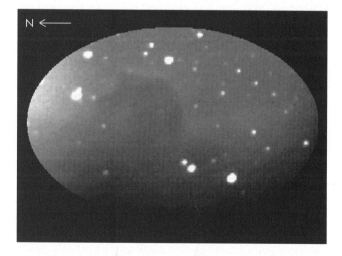

Figure 8.2 The Horsehead nebula, imaged with a CCD camera.

between the two dark nebulae; of 8th magnitude, it is distinctly nebulous, being surrounded by the reflection nebula NGC 2023.)

The North American Nebula RA_{2000}: 20 h 58.8 m Dec_{2000}: +44°20′ (see Figs 8.3 and 8.4)

This is an example of a dark nebula seen against a rich background of stars. Its catalogue numbers are NGC 7000 and C20. The obscuration in the "Bay of Mexico" region is very apparent, the star density dropping abruptly until only a few bright stars can be seen. Again this is easy to photograph and even the naked eye seems to show it as a brighter patch in the Milky Way to the south-east of Deneb (alpha Cygni).

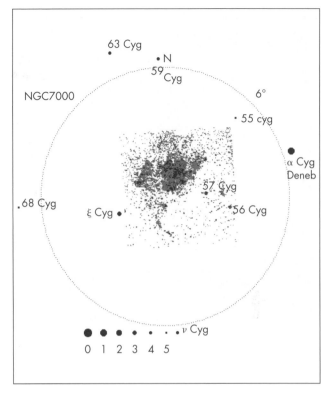

Figure 8.4   The position of the North American nebula.

Figure 8.3   The North American nebula, photographed with a 200 mm Schmidt camera.

8.4 Reflection Nebulae

8.4.1 Introduction

A reflection nebula (see Fig. 8.5, *overleaf*) appears bright and derives its name from the fact that it is visible by virtue of the reflected light of nearby stars. You can easily produce something similar in your own living room. Next time you are set up for a slide show, get out a duster, do a minute's housework on a corner that hasn't been touched for a while, and then shake the duster in front of the projector. Unless you are particularly houseproud, a beautiful bright reflection nebula will appear in the light beam. This is produced by the brightly illuminated dust particles reflecting a small part of the light back to you. The dust in gas clouds is not too dissimilar from the debris in your living room and behaves in exactly the same way when caught in the bright light from a star. The light quality is

Figure 8.5 A reflection nebula produced by dust shaken into the beam from a projector.

similar to that of the star, but the typical dimensions of the dust (around 100 nanometres) mean that it is better at reflecting blue wavelengths than red, so the light is made somewhat bluer. Of course, this means that when looking *through* such a nebula towards the star, some blue light is missing and the starlight appears redder. This interstellar reddening is well studied and forms the basis of one technique for distance measurement. This blue tinge can be seen often in the colour photographs taken with large Schmidt cameras, appearing in striking contrast to the bright red of the emission nebulae.

8.4.2 Nebulae

M78 RA$_{2000}$: 05 h 46.8 m   Dec$_{2000}$: +00°03′ (see Figs 8.6 and 8.7)

M78 (NGC 2068) is to be found in the Orion nebular complex, to the north-east of zeta Orionis. It is bright and easily seen in all telescopes. The appearance is somewhat reminiscent of a comet lacking a tail, but having a double nucleus, as there are two 10–11 magnitude stars less than 1′ apart visible inside its boundary. The whole of this part of the constellation is involved with dark nebulosity and so there are few background stars, quite different from just 3° or 4° away in the region of Orion's belt.

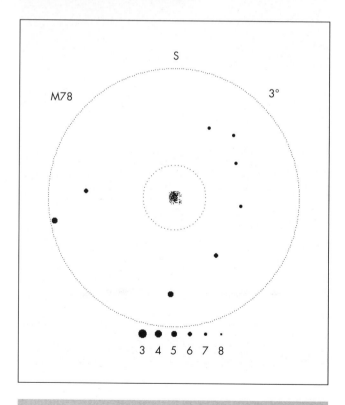

Figure 8.6   Finder chart for M78.

Figure 8.7 The visual appearance of M78 through 3-inch (75 mm), 6-inch (150 mm) and 12-inch (0.3 m) telescopes.

Hubble's Variable Nebula   RA$_{2000}$: 06 h 39.1 m   Dec$_{2000}$: +08°43′ (see Fig. 8.8 and Fig. 8.9, *overleaf*)

This interesting object, NGC 2261 or C46, is associated with the variable star R Monocerotis. Although quite small, it is easy to find as the nebulosity is bright. It looks like a tiny comet with R Mon as the nucleus and a tail 1′ long pointing to the north. Like the star, the nebula exhibits changes in brightness, but these are not due entirely to the star's variations, and shadowing by dust orbiting near the star is presumed to play a part. It is thought to be a young star surrounded by a remnant disk of gas and dust in its equatorial plane, so that starlight escapes to the surroundings primarily in cones round its poles. Other examples are seen in the form of bipolar nebulae having a bright lobe to each side of the star, but in the case of NGC 2261 we see only the one lobe. (Deep exposure photographs do reveal a second feature in the form of a narrow streak going south which bears a remarkable resemblance to a cometary anti-tail.) A 30 cm telescope under good conditions might reveal

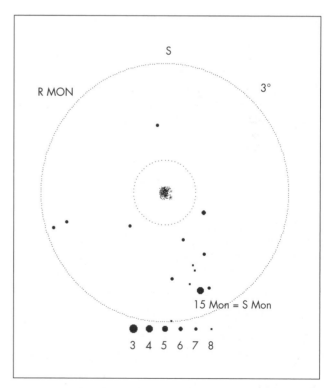

Figure 8.8 Finder chart for NGC 2261.

Figure 8.9 The visual appearance of NGC 2261 through 3-inch (75 mm), 6-inch (150 mm) and 12-inch (0.3 m) telescopes.

some of the internal structure in the nebula, but often it will appear featureless. There is, however, plenty of scope for photography or CCD imaging and these readily show the finer details. Views taken a month or more apart may well reveal changes in appearance.

8.5 Emission Nebulae

8.5.1 Introduction

An emission nebula is a short-lived end-product of the process of massive star formation. Radio observations show that deep inside the cold giant molecular clouds are found particularly dense "cores". These are identified as regions that have become dense enough that their self-gravity causes a gradual collapse. This contraction converts a considerable amount of gravitational energy into heat, which helps to stave off catastrophic collapse by raising the internal gas pressure. But this heat is slowly radiated from the core and ultimately the only situation that can halt the collapse is the central temperature and pressure becoming high enough for nuclear reactions to begin. The precise manner in which this happens is still somewhat unclear, but as the core collapses, conservation of angular momentum must cause it to spin faster and this results in much of the material being shed from the protostar to form an enveloping cocoon. This is dense enough to prevent the direct escape of light so that the newly formed star cannot be seen at this stage, though the residual warmth of the cocoon is detectable as an infra-red source. If the amount of material in the central star is a few times the mass of the Sun, then the nuclear reactions generate enough heat for its surface to shine brightly, emitting plenty of ultra-violet light. The ultra-violet radiation gradually evaporates the dust of the cocoon until it is dissipated entirely. This stage of star formation is thought to have been seen in the Eagle and Orion nebulae. Once all the immediate material has gone the star is free to wreak havoc on the less dense surroundings and, as stars tend to form in clusters rather than individually, it finds itself with many allies. The energetic ultra-violet light quickly evaporates the dust and ionises the gas so it shines with an emission spectrum, becoming an H II region. An ever-growing ragged bubble is blown out of the surroundings of the young cluster. Yet all this drama is hidden from view by the remaining dust until the cavity reaches the edge of the cloud. Then visible light can at last escape and the nebula becomes apparent to observers outside (see Fig. 8.10). Brighter streaks betray denser patches of material and the most resistant clumps and foreground dust appear in silhouette as dark nebulae. In other places the light can catch the remnants of dust to form a reflection nebula. In a short time the new cluster will have entirely devastated its surroundings and burnt its way through to become completely visible. In this haphazard way the dark clouds of the Galaxy are briefly punctuated by a succession of emission nebulae. We can see similar objects beautifully displayed in other spiral galaxies oriented face-on towards us.

8.5.2 The Spectra of Emission Nebulae and Planetary Nebulae

The maximum density that is reached in the interstellar medium, other than in and near protostars, is about 100 million atoms per cubic metre. This may sound high, but compared with the atmosphere surrounding you is unimaginably rarefied, as air density is more than 10 million million million million particles per cubic metre. The atoms in a nebula behave as individuals and this affects the spectrum of the light they emit. In a star the atoms are continually jostling and disturbing each other.

Figure 8.10 The Lagoon and Trifid nebulae (*left*) and the Swan and Eagle nebulae (*right*), photographed in red light with a 200 mm Schmidt camera. These are prominent objects in the southern part of the Milky Way.

Such jostling makes them give off a continuous spectrum with all wavelengths present, and inevitably this is also the type of spectrum from a reflection nebula. In an emission nebula the atoms become excited by absorbing ultra-violet light from the new stars and then can de-excite, producing their own light by the process of fluorescence. Since the atoms are free of disturbance from neighbours they emit a simple spectrum containing a limited set of wavelengths (see Fig. 8.11). The specific wavelengths emitted are characteristic of the particular type of atom, and enable the chemical composition of the interstellar medium to be investigated. In the visible region the strongest emission is from the hydrogen-alpha line at 656 nm, which gives the characteristic red colour in photographs of these objects. Visual observers, however, have difficulty detecting this since the eye is more sensitive to light in the green part of the spectrum. Another strong hydrogen line, H-beta at

486 nm, is to be found in the blue-green but there are also others, whose existence puzzled theorists for a long time. The hydrogen lines (and many others due to other chemical elements) can be produced easily from laboratory apparatus even though the gas density inside it is relatively high. The strong green lines from nebulae at 496 nm and 501 nm, however, have never been detected in the laboratory. It turns out that it is only the exceptionally low density in the gas clouds which allows the atoms to emit strongly in these lines. The two mentioned are produced by oxygen and there are many others from other elements. In terrestrial experiments they *are* produced but are overwhelmed by other emissions; because of this they are termed *forbidden* lines. In general, the source of a particular line is denoted by the chemical symbol for the element followed by a Roman numeral indicating its state of ionisation. If the line is forbidden the whole of this is enclosed in square brackets. Thus, for example, the oxygen lines at 496 nm and 501 nm are designated [O III].

The intimate relationship between star formation and the creation of emission nebulae means that they are concentrated in the Galactic plane, and thus more are visible looking towards the Galactic centre. All the brightest examples lie to the south of the celestial equator, putting temperate zone northern astronomers at a disadvantage.

8.5.3 Nebulae

The Great Orion Nebula   RA$_{2000}$: 05 h 35.4 m   Dec$_{2000}$: –05°23' (see Fig. 8.12)

M42 (NGC 1976) is a nebular showpiece accessible to northern and southern observers alike. It is plainly visible to the naked eye as a fuzzy patch below Orion's belt (see Fig. 8.1). Training binoculars on this patch will show a misty green haze extending over an area nearly the size of the Moon. With the smallest telescope the illuminating stars at the centre, a close group of four called the Trapezium, become resolvable. The brightest nebulosity is to be found close by, and a wedge of obscuring material extending to the east is prominent. This creates a strong overall impression of looking at the profile of a fish with its mouth open. A line of three stars shines brightly to the south of the Trapezium.

With a larger instrument, strands of the nebula curve away from either side of the dark intrusion, the brighter to the south-east with a fainter one to the north-west. The brightest nebulosity takes on a curdled appearance and has a somewhat square shape. In a 30 cm telescope the curdling is clearly resolved into a mass of small patches

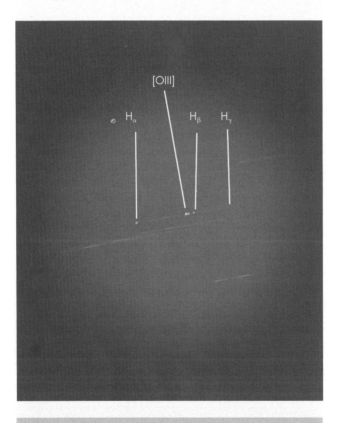

Figure 8.11   The spectrum of the planetary nebula NGC 6543 in Draco (the Cat's Eye Nebula, C6), made using an objective prism (see Section 11.9.2). Each strong emission line produces an image of the 20 arc second diameter nebula. Note, in contrast, the continuous spectrum produced by the central star.

Figure 8.12 The visual appearance of M42 through 3-inch (75 mm), 6-inch (150 mm) and 12-inch (0.3 m) telescopes.

and streaks, the south-east streamer curves further round and details in the south-west portion of the nebula appear fleetingly. In excellent conditions the prominent straight bar forming the south-eastern edge to the bright inner nebula shows tinges of red. At high power, two more stars can be perceived in the Trapezium. M42 responds well to hazy but steady skies, and to see it under these conditions is to understand precisely what Messier was referring to when, on publishing a drawing he said: "I have examined this nebula … in an entirely *serene* sky". It is an object to which you can return again and again.

Immediately to the north lies M43 (NGC 1982), a comma-shaped object on photographs, looking like a nebulous star at first glance and 1° north the reflection nebula NGC 1977 may be glimpsed surrounding 42 and 45 Ori (see Fig. 8.1).

The Lagoon Nebula   RA$_{2000}$: 18 h 03.7 m   Dec$_{2000}$: –24°20′ (see Fig. 8.13 and Fig. 8.14, *overleaf*)

Like M42, the Lagoon (M8, NGC 6523) is visible to the naked eye. In a telescope it can be found by pointing at 9 Sgr. The correct location can be verified by the presence of 7 Sgr, 15′ to the west of 9 Sgr, and a star cluster to its east. The brightest patch of the nebula is easily apparent near 9 Sgr and luminosity extends over the area covered by the

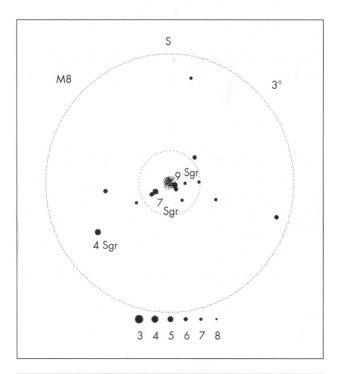

Figure 8.13 Finder chart for the Lagoon nebula.

Figure 8.14 The visual appearance of M8 through 3-inch (75 mm), 6-inch (150 mm) and 12-inch (0.3 m) telescopes.

cluster (which bears its own designation of NGC 6530). The presence of this nebulosity is most clearly signalled by the detection of the dark bar that forms the "lagoon". This runs obliquely between the bright patch and the cluster.

The Trifid Nebula   RA$_{2000}$: 18 h 01.9 m   Dec$_{2000}$: –23°02′ (see Figs 8.15 and 8.16)

M20, the Trifid nebula (NGC 6514), lies a little to the north of M8 (see Fig. 8.10). Its position is marked by a pair of 7/8th magnitude stars lying north–south and 8′ apart. The southern star is readily seen to be double and the dark lanes splitting up the nebula into the three parts for which it is named pass close by it. These show up more obviously than the nebulosity itself, which is not very bright. If you do not have dark skies, then a light pollution filter will probably be necessary to see anything of this object.

The Omega, Horseshoe or Swan Nebula   RA$_{2000}$: 18 h 20.9 m Dec$_{2000}$: –16°11′ (see Fig. 8.17 and Fig. 8.18, *overleaf*)

This is a bright object lying in the Milky Way in Sagittarius, carrying the Messier designation M17 and appearing in the *New General Catalogue* as NGC 6618. It is clearly visible in finder telescopes and unmistakable in the main telescope, the general shape being very reminiscent of a floating swan. In smaller telescopes, the body is seen to be of variable

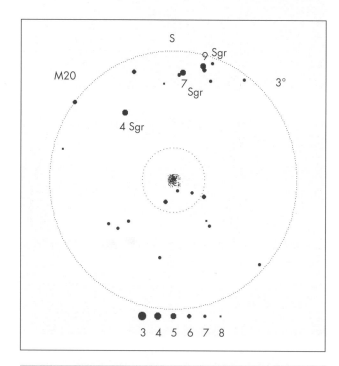

Figure 8.15 Finder chart for the Trifid nebula.

Figure 8.16 The visual appearance of M20 through 3-inch (75 mm), 6-inch (150 mm) and 12-inch (0.3 m) telescopes.

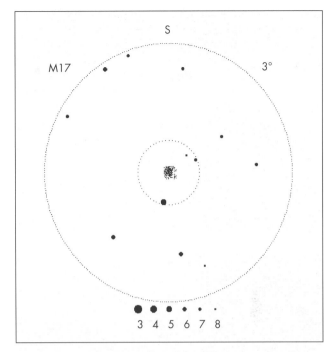

Figure 8.17 Finder chart for M17.

brightness along its length, and with increasing aperture more and more features can be seen crossing it. The loop forming the neck and head also reveals detail and further fainter nebulosity is visible to the east of it. There is a star cluster associated, which covers an area about 20′ across and lies mostly to the north of the nebula.

The Eagle Nebula   RA$_{2000}$: 18 h 18.8 m   Dec$_{2000}$: −13°47′ (see Figs 8.19 and 8.20, *overleaf*)

M16 (nebula IC 4703 + cluster NGC 6611) lies $2\frac{1}{2}°$ away from M17, in Serpens (see Fig. 8.10). Unlike its neighbour, the view is dominated by the star cluster, with up to 50 stars visible in larger apertures. The rather faint nebulosity lies throughout the cluster and extends beyond it to the south. The area photographs well and there are seen to be many dust intrusions of all shapes and sizes. The biggest of these, a triangular wedge protruding from the northern boundary, can be seen in a 20 cm or larger telescope. For a challenge you might like to try observing at high power to see if you can catch a glimpse of the dark lanes right at the centre (the aforementioned wedge points straight towards them). Two brighter stars are situated within them to make it more awkward. These lanes were imaged in intricate detail by the Hubble Space Telescope and seem to show that here more stars are waiting to emerge from

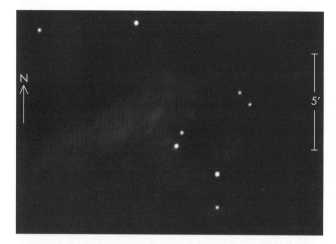

Figure 8.18 The visual appearance of M17 through 3-inch (75 mm), 6-inch (150 mm) and 12-inch (0.3 m) telescopes.

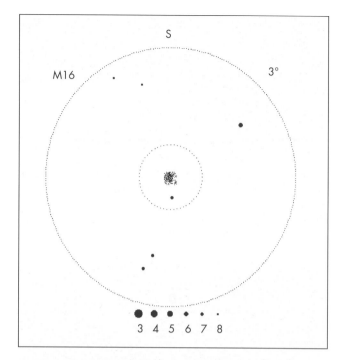

Figure 8.19 Finder chart for M16.

their cocoons. Meanwhile, the visible cluster stars are evaporating their less dense surroundings, in some cases leaving the newer cocoons isolated.

8.6 Supernova Remnants

8.6.1 Introduction

Every now and then the familiar pattern of the starry sky is disrupted by the appearance of a new star outshining all others: a supernova. This represents the last fling of a massive star as its centre runs out of energy and collapses into a neutron star. The outer layers are thrown off as material involved in the collapse hits the new central object at high speed and rebounds, setting off a powerful shock-wave that sweeps all before it. The associated heating produces a final burst of nuclear reactions in the outer material, creating a range of heavy elements. The expansion of the ejected hot material produces a rapid huge rise in luminosity of the star, and the radioactive decay of some of the new elements keeps it shining for many months, gradually fading all the while. Finally the expanding ejecta

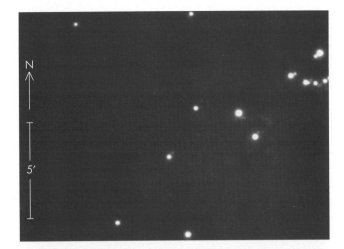

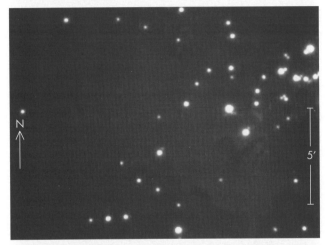

Figure 8.20 The visual appearance of M16 through 3-inch (75 mm), 6-inch (150 mm) and 12-inch (0.3 m) telescopes.

grow to a visible dimension and may be kept glowing by energy supplied from the central neutron star for several hundred years. As the shell ploughs into the surrounding interstellar medium its edges become denser and are compressed into filaments. One important role that this process plays is to enrich the interstellar gas with heavy elements, and this is the only known way in which atoms other than hydrogen and helium can be distributed. Another side-effect is that the shock travelling outward into the interstellar medium can set off the collapse of denser clouds into new stars. A whole chain of star formation regions can be set up. The average rate of supernovae in spiral galaxies seems to be a few per century, yet none has been seen in our own galaxy for nearly 400 years. Presumably, by bad luck, all those that have occurred have been hidden from view by intervening dust clouds. The best we have had was a supernova in the nearby Large Magellanic Cloud in 1987. This was well-studied optically, and in addition a few of the vast number of neutrinos that it emitted were detected by special instruments that have recently been set up around the world. These are mainly used for studying neutrinos from the Sun but will detect any future supernova in the Galaxy even if we cannot see it. Hence we should soon be able to tell if the Milky Way also has a rate of a few supernovae per century. Tracking down the remnants is, in the meantime, a worthwhile observing goal. Bear in mind that many of the atoms in your own body were created in a supernova and sprinkled into space in such a remnant before eventually becoming part of the Solar System. You can hardly get closer to the Universe than this.

8.6.2 Nebulae

The Crab Nebula   RA$_{2000}$: 05 h 34.5 m   Dec$_{2000}$: + 22°01′ (see Figs 8.21 and 8.22, *overleaf*)

It is fitting that this object, M1 or NGC 1952, heads Messier's list as it has single-handedly brought us an understanding of more celestial phenomena than anything else. It is the result of a supernova that was seen in 1054. The ejecta has since spread out to an angular size of 6′ by 4′. It is visible in any amateur telescope as an amorphous elongated patch with no structure. In a 30 cm telescope the "bays" intruding into the shape should be seen and the uneven brightness over the surface of the nebula be apparent. The neutron star that powers the nebula is rotating very rapidly and flashes 30 times a second in time with the rotation. Its average brightness is 16th magnitude and it lies close to an unrelated star of similar appearance. The pair lie beyond the capability of small telescopes, at least

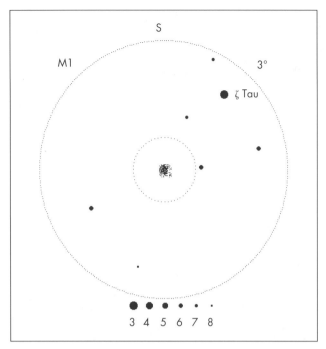

Figure 8.21 Finder chart for the Crab nebula.

for visual observations. Photography or imaging easily reveal them and also show a network of emission filaments superposed over the general synchrotron emission.

The Veil Nebula or Cygnus Loop (see Figs 8.23 and 8.24)

Eastern   RA$_{2000}$: 20 h 56.4 m   Dec$_{2000}$: +31°43′
Western   RA$_{2000}$: 20 h 45.7 m   Dec$_{2000}$: +30°43′

The supernova explosion that produced this remnant took place an estimated 15 000 years ago. Coupled with a smaller distance (2600 as against 6500 light years) this makes the angular size of this nebula much bigger than the Crab and it is not possible to view it in its entirety in the telescope. The central star (whose identity has not been established) ceased to power synchrotron emission from the gas long ago and what we see now is the ejected material continuing to plough into the interstellar medium. The speed is fast enough (45 km s^{-1}) to maintain a shock wave that heats the gas and keeps it emitting. This outer shell is now very patchy and only parts of the complete loop are visible, within a 3° circle. The brightest part, C33 or NGC 6992, NGC 6995 and IC 1340, is the north-eastern side; to the south-west is C34 (NGC 6960) and there are fainter patches in between. With binoculars the eastern part can be glimpsed as a faintly glowing curve,

Figure 8.22 The visual appearance of M1 through 3-inch (75 mm), 6-inch (150 mm) and 12-inch (0.3 m) telescopes.

Figure 8.23 The Veil nebula, photographed in red light with a 200 mm Schmidt camera.

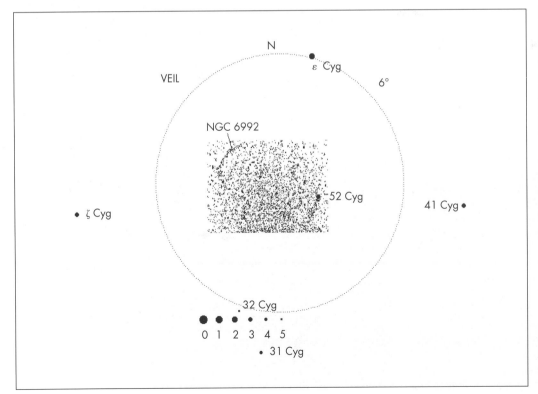

Figure 8.24 The position of the Veil nebula.

and telescopes of increasing size reveal more details. The field containing the best part of the northernmost section (NGC 6992) is marked by a 10th magnitude stellar triangle. Its longest side is 2.5′ north–south and the point is towards the west. In good conditions, with reasonable seeing, the general patchy glow is readily perceived and finer details momentarily flash into view. In poorer skies a nebular filter helps considerably. The western part passes next to 52 Cyg. In the telescope this can be seen as a narrowing filament extending north of the star. To the south of 52 Cyg it grows considerably fainter and is difficult to detect.

8.7 Planetary Nebulae

8.7.1 Introduction

While supernovae remnants represent the endpoints of the impetuous lives of massive stars, the planetary nebulae are consequences of the gentler ageing processes of more modest objects. Stars similar to the Sun will eventually increase in luminosity and size as the internal nuclear reactions modify their chemical composition. The central core on which the star has relied for a relatively stable rate of energy production during its life gradually becomes clogged with helium. The rate at which hydrogen nuclei encounter each another declines and so the conversion to helium slows down. The resultant drop in energy production lowers the pressure, causing the core to contract. This releases gravitational energy (as when the star was first formed), reheating the core. Slowly the core temperature rises until the helium nuclei start to collide violently enough to react, burning to carbon. The heat from this new process augments the conversion rate of hydrogen to helium which continues in the encircling outer zone. The star expands in response to the vastly increased energy flowing out from the conflated volume undergoing reactions. It becomes so big that the surface can dissipate all this energy at the relatively low temperature of 3000–4000 K, so it becomes redder in colour and is now termed a *red giant* star. An expanding unstable outer envelope develops which eventually reaches the local escape velocity and so becomes free from the rest of the star. Ultimately the star runs out of both hydrogen and helium, and undergoes a final contraction and associated rise in temperature. It ends up as a very hot and rather small *white dwarf* star. The surface temperature of this can be as high as 200 000 K so its light emission is almost entirely in the ultra-violet part of the spectrum. This causes the tenuous gas in the escaped envelope to fluoresce in the same way as that in emission nebulae and produces a similar spectrum (see Section 8.5.2). The resultant bright nebula is termed a *planetary nebula* (the nomenclature dates back to the days when the nature of these nebulae was entirely obscure). The shapes that these can take are rather varied but are presumed to be influenced by two main factors: (i) the original envelope may have been far from spherical when it detached from the star, and (ii) the star probably has a strong stellar wind flowing from its surface which can help to gather up the envelope into a more regular shell. As the central white dwarf radiates away its stored internal heat, its temperature drops so less ultra-violet light is emitted and this in combination with the continued gradual expansion of the nebula causes it to fade gradually away.

Figure 8.25
Images of the Ring nebula taken through red (*left*) and green (*right*) filters. Note the bright rim in the red image, and the slightly smaller overall size and brighter centre of the green image.

There are several planetary nebulae that have a high surface brightness and so are easily observable, and in fact of all objects these are the ones whose visual appearance most resembles the photographs seen in books.

8.7.2 Nebulae

The Ring Nebula   RA$_{2000}$: 18 h 53.6 m   Dec$_{2000}$: +33°03′ (see Figs 8.25, 8.26 and 8.27)

This is probably the easiest of all the planetaries, being bright, of reasonable size and lying in a position conveniently marked by the prominent stars beta and gamma Lyrae. Messier designated it M57 and its NGC number is 6720. It is picked up in any telescope as a grey oval at first sight, which readily changes to an oval ring on slightly closer examination. It takes magnification well and the larger image makes it easier to note that the interior of the ring is also glowing, being noticeably less dark than the sky. The overall size is 80″ × 60″ with the centre half these dimensions. A 15 cm telescope will show the 12th magnitude star close to the eastern edge, while larger apertures show that the ends of the oval are less bright

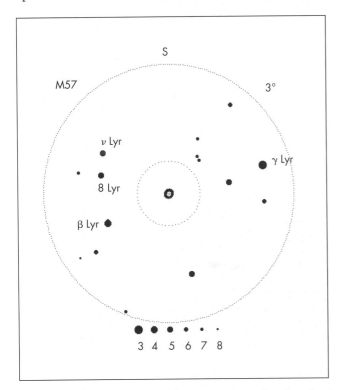

Figure 8.26 Finder chart for the Ring nebula.

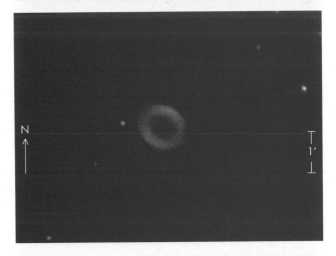

Figure 8.27 The visual appearance of M57 through 3-inch (75 mm), 6-inch (150 mm) and 12-inch (0.3 m) telescopes.

than its sides. The Ring is a good subject on which to experiment with colour imaging, as the physical conditions within it mean that the principal emission lines in the red and green arise from different areas. This leads to a noticeable difference in its appearances in images taken through filters isolating different colours (see Fig. 8.25).

The Dumbbell Nebula RA_{2000}: 19 h 59.6 m Dec_{2000}: +22°43' (see Figs 8.28 and 8.29)

M27 (NGC 6853) is another bright object and can be found 3° almost due north of gamma Sagittae. It can be missed at first, but once located is unmistakable with its two touching lobes. (Do not be misled by the name: it is much closer to a hourglass in shape.) The southern part can be seen to be brighter and the central star is discernible with a 30 cm telescope. In good conditions there is some trace of nebulosity outside the main dumbbell region, extending to 7' or 8' across. The "height" of the cones is 4'. Images show that the bright cones have a mottled structure with scale of about 20", but this does not seem to be readily apparent to visual inspection.

The Saturn Nebula RA_{2000}: 21 h 04.2 m Dec_{2000}: −11°22' (see Figs 8.30 and 8.31)

This bright planetary (NGC 7009, C55) lies in Aquarius. It

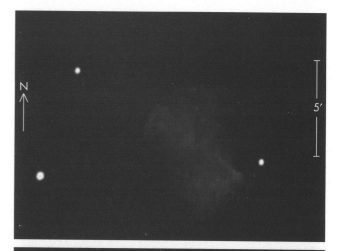

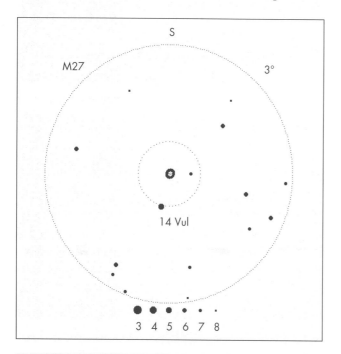

Figure 8.28 Finder chart for the Dumbbell nebula.

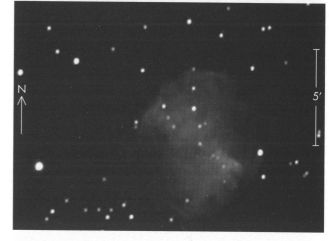

Figure 8.29 The visual appearance of M27 through 3-inch (75 mm), 6-inch (150 mm) and 12-inch (0.3 m) telescopes.

is of sufficient size ($40'' \times 20''$) that its nature is easily apparent and it first appears as a greenish-blue disk, brighter towards the centre. With a moderate telescope the shape is seen to be oval and there are thin extensions at its ends, though these will be difficult without a filter from locations with significant light pollution.

The Helix Nebula   RA$_{2000}$: 22 h 29.6 m   Dec$_{2000}$: $-20°48'$
(see Figs 8.32 and 8.33, *overleaf*)

NGC 7293 or C63 in Aquarius has fully half the angular diameter of the Moon. Its name stems from its photographic appearance and visually it is simply a faint ring. It is brighter at the south-west and north quadrants where the helical strands overlap. To help pin down the position, the western side is close by a triangle whose longest side (12') is north–south; the short side of 3' at the southern end points east and has a double star (20'' separation) at this vertex. In line with the first side there is a brighter star 21' beyond to the north, and this star has a slightly fainter one 13' to its west. Under good conditions the Helix can be seen in binoculars or finder telescopes. The central hole can be noted with apertures of 20 cm or more.

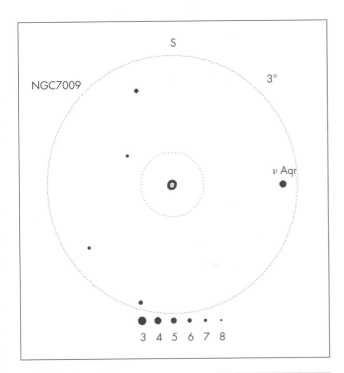

Figure 8.30   Finder chart for the Saturn nebula.

Figure 8.31   The visual appearance of NGC 7009 through 3-inch (75 mm), 6-inch (150 mm) and 12-inch (0.3 m) telescopes.

Figure 8.32 The Helix nebula, photographed in red light with a 200 mm Schmidt camera.

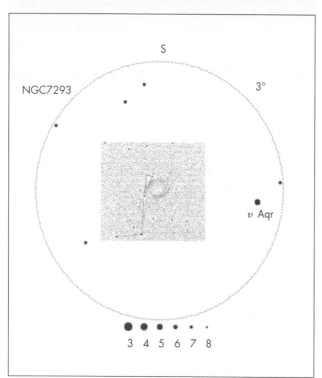

Figure 8.33 Finder chart for the Helix nebula.

Chapter 9

Galaxies

9.1 Introduction

Most objects to be seen in the sky are either pin-points of light (stars), or fuzzy blobs. The fuzzy blobs were not understood at all until a little over a century ago. Were they aggregations of stars so far away that the individual members could not be separated from each other, or were they truly amorphous objects? As bigger telescopes were built, direct observations were able to distinguish stars in some of these objects, but other blobs remained unresolved. Then in 1864, William Huggins put a spectroscope in place of the eyepiece of his telescope, and the problem was resolved. Some of the blobs turned out to have an emission line spectrum (see Fig. 9.1), and so must be masses of hot gas. Others had an absorption line spectrum (see Fig. 9.1) like that of stars, and so must be aggregations of very distant stars. The first of these groups we now know as the nebulae (see Chapter 8), and it includes things like H II regions, Planetary Nebulae, Supernova Remnants etc. Among the second group, many were found to be groups of stars within our own Milky Way galaxy, such as the galactic clusters (Chapter 7), or groups just outside our galaxy, and orbiting around it (globular clusters – Chapter 7).

Some of these "stellar nebulae", however, obstinately refused to show their stars, and many of this latter group had a spiral shape when seen in large telescopes. Eventually, well into this century when the 100-inch (2.5 m) Hooker telescope became available, one of these objects, known as M31, was resolved into stars. It then became clear why it was so difficult to see the stars, because the object was so much further away than anything observed

Emission line spectrum

Absorption line spectrum

Figure 9.1 Emission and absorption line spectra.

up to that time – about 2 930 000 light years (930 kpc). At first M31 and similar objects were called extra galactic nebulae, but nowadays we know them as galaxies. M31 is among the closest of the galaxies, and is a spiral seen almost edge-on (see Fig. 9.2, *overleaf*). It is frequently called the "Great Galaxy in Andromeda", and also NGC 224 (after the *New General Catalogue of Nebulae and Clusters of Stars*).

Galaxies range in size from about 10 000 000 (10^7) times the mass of the Sun to 1 000 000 000 000 (10^{12}) times the mass of the Sun. As well as stars, many galaxies also contain large amounts of gas and dust. The latter may sometimes be seen as dark "lanes" in some galaxies where

Figure 9.2 The Great Galaxy in Andromeda (M31, NGC 224).

the dust is obscuring the more distant stars (see Fig. 9.3). Also some galaxies may have black holes with masses from 1000 to 100 000 000 times the mass of the Sun at their centres (Section 9.5).

Today, galaxies form one of the leading areas of research in astronomy, and as well as being of considerable interest in their own right, they can tell us much

Figure 9.3 The galaxy NGC 891 (C23), an edge-on spiral galaxy showing a prominent absorption lane due to interstellar dust clouds.

about the early stages in the life of the Universe as a whole (cosmology). From observations using spectroscopes, we find that we are moving towards the Andromeda galaxy at a speed of some 275 km s^{-1}. Most of this, however, is due to our own orbital motion around the Milky Way galaxy. Even if it were a straight-line velocity, it would take 2 000 000 000 years before we were in any danger of there being a collision, so there is no need to worry!

Observations of other galaxies by Edwin Hubble in the late 1920s showed that, in general, they were moving away from us. Furthermore, he found that the further away a galaxy was, the faster it was moving away from us. This phenomenon is now known as the *redshift* of the galaxies, because optical spectrum lines are Doppler-shifted towards the red end of the spectrum by the motion. It is fundamental to our ideas on how the Universe came into being, with most theories today suggesting that the red-shift is the remnant of an explosion (Big Bang) occurring between 8 000 000 000 and 20 000 000 000 years ago, at which time all the material and objects we now see in the Universe were compressed into a very very tiny region indeed.

Although many galaxies showed a spiral form, this was not true of all galaxies, and Hubble also devised a classification system that is still in use today. The three main groups were spiral galaxies, elliptical galaxies and irregular galaxies (see Fig. 9.4). Many of the latter type of galaxy contain regions of intense activity and energy emission, and along with Seyfert galaxies, quasars, etc. (see below), are called *active galaxies*. The energy sources for the centres of active galaxies, and also for some of the more "normal" galaxies, are generally thought to be material in the form of stars, planets and gas clouds falling into super-massive black holes. The Hubble classification of galaxies is shown in Fig. 9.5 (*overleaf*).

The Andromeda galaxy and our own slightly smaller Milky Way galaxy are both of type Sb.

9.2 Spiral Galaxies

Spiral galaxies are the most interesting group among the galaxies in terms of their visual appearance. Hubble distinguished two main sub-classes, each of which was then subdivided into three on the basis of how tightly wound were the arms (see Fig. 9.5, *overleaf*). The two sub-classes are the normal spirals, which have the spiral arms emerging from the nucleus directly, and the barred spirals, where the arms originate from a linear extension to the nucleus (or bar).

M74

M32

M82

Figure 9.4 Examples of spiral (M74), elliptical (M32) and irregular (M82) galaxies.

The nucleus of a spiral galaxy is usually a somewhat flattened sphere up to 20 000 or 30 000 light years (6 to 10 kpc) across. The stars comprising the nucleus are generally low-mass, older stars. Their temperatures are comparable with that of the Sun or less, and so the nucleus tends to have a reddish tinge to it on images taken with very long exposures. Generally there is very little gas and dust to be found within the nucleus. The nuclei of some spiral galaxies have a very intense and tiny central region. In extreme cases, this central region may emit several times the energy coming from the whole of the rest of the galaxy. Such galaxies are known as Seyfert galaxies, or in really extreme cases, as quasars (Section 9.5). Even comparatively normal spiral galaxies, like our own Milky Way galaxy, however, may exhibit the same phenomenon on a much reduced scale. It is widely thought that these intense central regions contain a black hole with a mass from 1000 to 100 000 000 times that of the Sun. Matter from the rest of the galaxy is in the process of being absorbed by the black hole, and energy is emitted in large amounts as that material spirals inwards. This energy then provides the power for the central emission region that we observe.

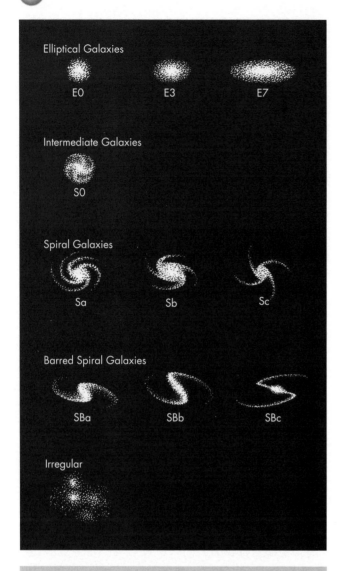

The Milky Way itself is our own galaxy, as we see it from an internal view point. The dust in our own galaxy limits our view of it to within a few thousand light years from the Sun, and dust absorption also causes the patchy, irregular appearance of the Milky Way (Chapter 10). How the spiral arms are formed is still a matter for debate. They may be due to tidal interactions as two galaxies undergo a near collision and/or they may arise from pressure waves emanating from explosions at the centre of the galaxy, perhaps as larger quantities of matter than usual fall into the central black hole.

9.2.1 Observing Spiral Galaxies

Galaxies in general are among the most difficult objects to observe, for despite their huge sizes and enormous luminosities, they are so far away that even the best examples are very faint. A low-power eyepiece is thus essential to make best use of the available light, and the shorter the focal ratio of the telescope, the better. Low-power binoculars (such as 7×50 or so) may be even better than a telescope. The spiral arms themselves usually require telescopes with diameters of 20 inches (0.5 m) or more and very dark clear skies, if they are to be seen visually. So smaller telescopes will generally only show the nucleus. With few exceptions, it will be necessary to observe on nights when the Moon is absent from the sky, and to allow the eyes to become fully dark adapted (see Chapter 2). Averted vision (Chapter 2) is also likely to be helpful. In addition, the observer may find it useful to move the telescope very slightly back and forth once the galaxy is in the field of view. The moving image thus produced will often show up better than a stationary one.

The tabulated magnitude (brightness) of a galaxy is not automatically a guide to the ease with which it may be seen, as it would be with a star. This is because it is the integrated magnitude; that is to say, the magnitude obtained using the total light received from the whole area covered by the galaxy. An apparently bright galaxy may thus actually be very difficult to detect if it also covers a large area of sky. Conversely a faint but compact galaxy may be quite an easy object. Thus M74 and M82 (see below) both have integrated magnitudes of 8.8, but the latter is much easier to see. The situation is further complicated because some galaxies have a small bright nucleus, whereas others have a much more uniform intensity. The nucleus in the former can often

Figure 9.5 The Hubble classification of galaxies.

The spiral arms may be up to 200 000 light years (60 kpc) across. They have a bluish tinge on very long exposures, owing to the presence of numerous large hot stars. Such stars must also be young, and so, on a galactic time scale, the spiral arms are also young. The arms contain much gas and dust. The dust absorbs light very strongly, and so large dusty regions can sometimes appear as dark areas silhouetted against the rest of the galaxy (see Fig. 9.3). The Sun is well out into the spiral arms of the Milky Way galaxy, at some 30 000 light years (10 kpc) from the centre.

therefore be seen in quite small telescopes, where it will appear star-like, while the more uniform object of similar integrated magnitude would require a much larger instrument.

There are a large number of spiral galaxies to be observed (Appendix 4). We look below at some of the best and also at some more typical examples and at how you may expect to see them in various sizes of telescope.

M31 (Great Galaxy in Andromeda, NGC 224) RA$_{2000}$: 00 h 42.7 m Dec$_{2000}$: +41°16′
Visual Magnitude: 3.5; Size: 1° × 3°; Distance: 930 kpc (2.9×10^6 ly); Type: Sb

Only two spiral galaxies[1] are visible to the naked eye. One is our own, and we can see only a small part of it as the Milky Way. The other is the Great Galaxy in Andromeda, also known as M31. It can be seen on clear dark nights just to the west of ν And (see Fig. 9.6). It is easy to star hop from β And via μ and ν And to M31, so that it can be pointed out even to inexperienced observers. In less than ideal observing conditions, it is often still detectable using averted vision (Chapter 2).

The view through a telescope (see Fig. 9.7, *overleaf*) shows the central nucleus clearly. However, M31 is so large (over 3° or six times the size of the Full Moon along its long axis), that low-power, wide-angle eyepieces are needed on a short focal length telescope to see more than a small part of it at any one time. Binoculars can be used to advantage and will sometimes show parts of the disk surrounding the nucleus. In a telescope you may find that the disk can be detected by offsetting from the galaxy and then letting it drift back into the field of view. The disk will then appear as a brightening of the background before the nucleus drifts into the field of view. No further detail is to be seen visually by using higher magnifications or larger telescopes, though on long-exposure images, the dark dust lanes soon start to show up (see Fig. 9.7, *overleaf*).

M31 has several small satellite galaxies, two of which are quite easy to see (M32 and M110 – Section 9.3).

M51 (Whirlpool Galaxy, NGC 5194) RA$_{2000}$: 13 h 29.9 m Dec$_{2000}$: +47°12′
Visual Magnitude: 8.4; Size: 8′ × 11′; Distance: 4.6 Mpc (15×10^6 ly); Type: Sc

---

[1] The Large Magellanic Cloud (Section 9.4) may be a spiral galaxy in the process of formation. However, the spiral shape is very poorly developed (if it is there at all) and so here the LMC is classed with the irregular galaxies.

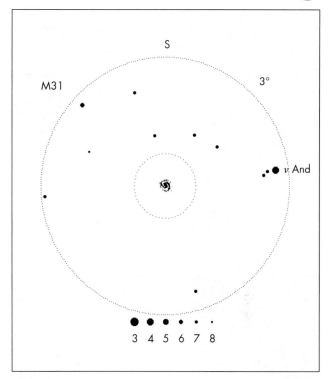

Figure 9.6 Finder chart for M31.

and

NGC 5195 RA$_{2000}$: 13 h 30.0 m Dec$_{2000}$: +47°16′
Visual Magnitude: 9.6; Size: 4′ × 5′; Distance: 4.6 Mpc (15×10^6 ly); Type: Irregular

To be found in Canes Venatici (see Figs 1.12, 1.13 and Fig. 9.8, *overleaf*), M51 is one of the most beautiful of spiral galaxies (see Figs 9.9 and 9.10, *overleaf*). It was the first galaxy in which spiral structure was seen. This was by Lord Rosse in 1845 using a 72-inch (1.8 m) telescope with a speculum metal mirror. With a modern telescope and on a good clear dark night, the spiral structure may be glimpsed in a 12-inch (0.3 m) telescope (see Fig. 9.10, *overleaf* (*bottom*)). Smaller telescopes will show the nuclei of both M51 and its companion NGC 5195. Although NGC 5195 is fainter than M51, its smaller size means that both nuclei are about equally visible. The companion is slightly more distant from us than M51, and the two galaxies probably had a close encounter some millions of years ago. An extension of a spiral arm from M51 seems to link to NGC 5195, but it actually overlies the companion.

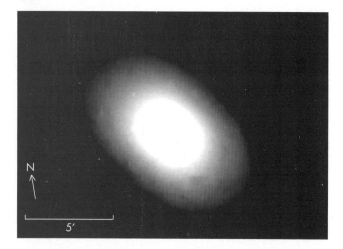

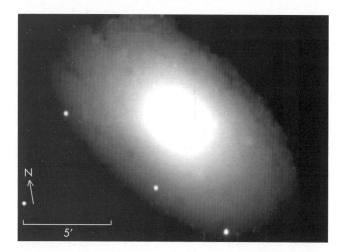

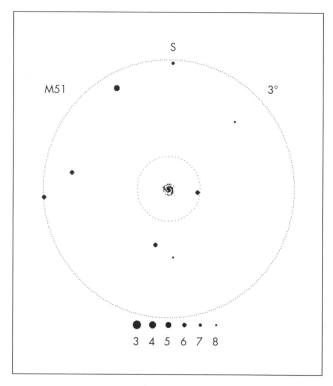

Figure 9.8 Finder chart for M51.

Figure 9.7 The visual appearance of the Great Galaxy in Andromeda through 3-inch (75 mm), 6-inch (150 mm) and 12-inch (0.3 m) telescopes. (Figure 9.2 shows a CCD image of the Great Galaxy in Andromeda through a 6-inch (150 mm) telescope using a 400 s exposure.)

Figure 9.9 A 10-minute exposure CCD image of M51.

M74 (NGC 628) RA$_{2000}$: 01 h 36.7 m Dec$_{2000}$: +15°47′
Visual Magnitude: 9.2; Size: 10′ × 10′; Distance: 7.8 Mpc
(26 × 10^6 ly); Type: Sc

This is reckoned by many to be the most difficult of the
Messier objects to observe, so you may put a feather in your
cap when you succeed! It is to be found in Pisces about 1$\frac{1}{4}$°
slightly to the north and to the east of η Psc (see Fig. 9.11,
overleaf), and just over the border from Aries. It is an Sc
type galaxy with multiple arms (see Fig. 9.4 (*top left*)). Its
visual magnitude is about 9, and this is mostly from the
nucleus, so that in small telescopes it may seem almost
star-like. It is marginally detectable in a 3-inch (75 mm)
telescope, but may certainly be seen in 6 to 8-inch (150 mm
to 200 mm) instruments under good observing conditions
(see Fig. 9.12, *overleaf*). Unlike many galaxies, the concen-
tration of brightness into the nucleus means that higher
magnifications may be better than lower ones. The nucleus
may then be seen as a small disk rather than as star-like.

M104 (Sombrero Galaxy, NGC 4594) RA$_{2000}$: 12 h 40.0 m
Dec$_{2000}$: −11°37′
Visual Magnitude: 8.3; Size: 4′ × 9′; Distance: 12 Mpc (40 ×
10^6 ly); Type: Sb

M104 (see Fig. 9.13, *overleaf*) is a large spiral galaxy belong-
ing to the Virgo cluster that is seen almost edge-on. Its dust
clouds, concentrated in the spiral arms, are seen as a dark
band crossing the nucleus. In small telescopes the galaxy is

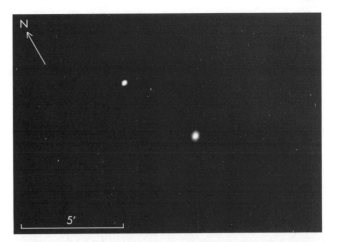

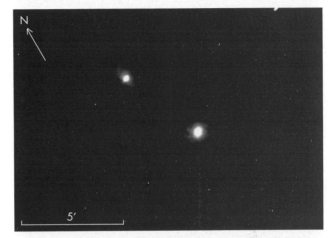

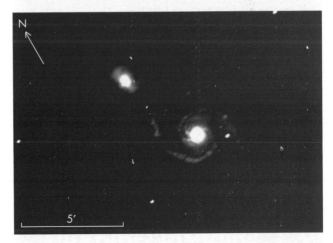

Figure 9.10 The visual appearance of M51 through 3-inch
(75 mm), 6-inch (150 mm) and 12-inch (0.3 m) telescopes.

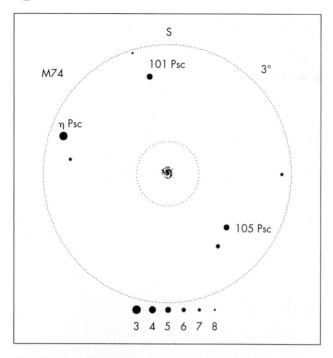

Figure 9.11   Finder chart for M74.

visible, but largely featureless. Telescopes with an aperture above about 6 inches (150 mm), however, will show the dark band quite clearly (see Fig. 9.14). The visual appearance, although similar to that on long-exposure images (see Fig. 5.12), is in fact only of the central part of the nucleus.

M83 (NGC 5236)   RA$_{2000}$: 13 h 37.0 m   Dec$_{2000}$: –29°52′
Visual Magnitude: 8.2; Size: 10′ × 11′; Distance: 10 Mpc (30×10^6 ly); Type: Sc or SBc

M83 (see Fig. 9.15, *overleaf*) in Hydra is a magnificent object for southern observers. In a small telescope only the central nucleus is detectable, but the spiral arms can be detected in a 12-inch (0.3 m) or perhaps slightly smaller telescope (see Fig. 9.16, *overleaf*). On long-exposure images (see Fig. 9.17, *overleaf*) it appears as a beautiful open spiral, with a hint of a bar through the nucleus. Knots of bright young stars show up and are the reason why the spiral arms can be seen in relatively small telescopes.

9.3   Elliptical Galaxies

Even on long-exposure images, elliptical galaxies do not have the visual interest of the spiral galaxies. They are aggregations of stars arranged in a more or less flattened

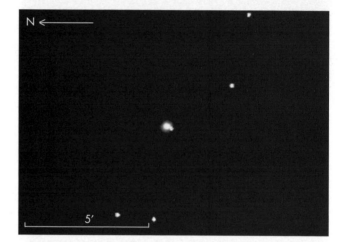

Figure 9.12   The visual appearance of M74 through 3-inch (75 mm), 6-inch (150 mm) and 12-inch (0.3 m) telescopes.

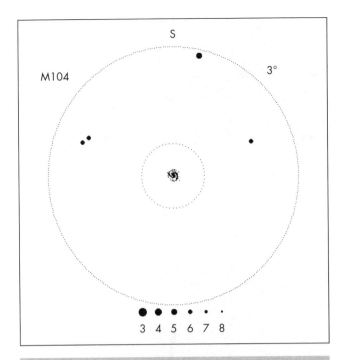

Figure 9.13 Finder chart for M104.

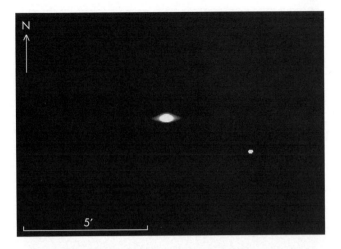

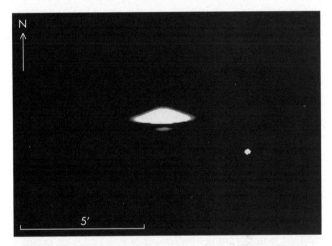

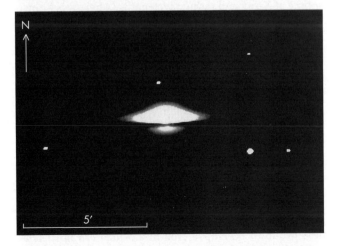

sphere. The stars are mostly smaller and older than those in the spiral arms of spiral galaxies, and so on long-exposure images, elliptical galaxies can have a reddish tinge. In this respect they resemble the nuclei of spiral galaxies. Some elliptical galaxies are among the largest known galaxies, while at the other end of the scale they can be as small as globular clusters (Chapter 7). Their shapes range from spherical (Class E0: Fig. 9.5) to being three times longer than they are wide (Class E7). A few elliptical galaxies, such as M87 (see below), are ejecting jets of material, sometimes at high velocities. These jets may be associated with super-massive black holes at the centres of the galaxies, and along with active galaxies (Section 9.5) form one of the leading areas of research in astronomy.

Through a small telescope elliptical galaxies will appear as a featureless fuzzy blob. It should, however, be possible to discern the shape of the galaxy. Some ellipticals are strongly centrally condensed, so they may appear star-like in small telescopes.

M87 (NGC 4486)   RA$_{2000}$: 12 h 30.8 m   Dec$_{2000}$: +12°24′
Visual Magnitude: 8.6; Size: 7′ × 7′; Distance: 13 Mpc (40 × 10^6 ly); Type: E1

M87 (see Fig. 9.18, *overleaf*) is one of the largest, and possibly the largest, of the known galaxies with a mass some 30 times that of the Milky Way galaxy. It also has

Figure 9.14 The visual appearance of M104 through 3-inch (75 mm), 6-inch (150 mm) and 12-inch (0.3 m) telescopes.

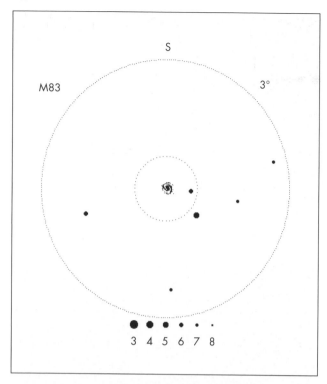

Figure 9.15   Finder chart for M83.

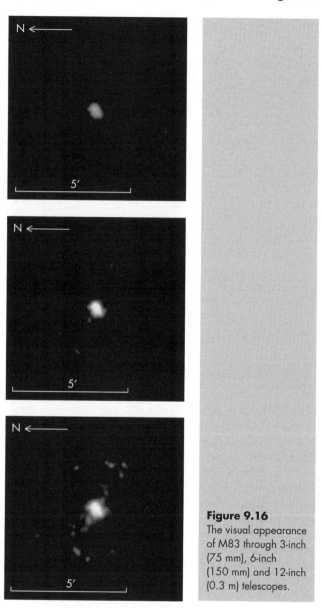

Figure 9.16
The visual appearance of M83 through 3-inch (75 mm), 6-inch (150 mm) and 12-inch (0.3 m) telescopes.

around 10 000 globular clusters orbiting around it. It is a member of the Virgo cluster of galaxies. The galaxy's nucleus is undergoing a violent upheaval and is an intense source at X-ray and radio wavelengths (where it is known as Virgo A and 3C274). It also has a high-velocity jet of material emerging to one side (see Fig. 9.19). The jet can be seen visually if you have a 100-inch (2.5 m) or larger telescope! In smaller instruments it may be seen as an almost circular image which fades uniformly from its centre towards the edge (see Fig. 9.20).

M110 (NGC 205)   RA$_{2000}$: 00 h 40.4 m   Dec$_{2000}$: +41°41′
Visual Magnitude: 8.0; Size: 10′ × 17′; Distance: 640 kpc (2.1×10^6 ly); Type; E6

This is one of several small satellite galaxies of the Great Galaxy in Andromeda, M31 (Section 9.2). M110 is shown in Figs 9.21, 9.22 and 9.23 (*overleaf*) along with another satellite galaxy, M32 (see also Fig. 9.4). It may be seen to be a pronounced ellipse even in a 3-inch (75 mm) telescope, but little further detail is to be found with larger telescopes.

9.4 Irregular Galaxies

Irregular galaxies are essentially all those that are not spirals or ellipticals. There is therefore a very great variation among them. Many of them have violent events occurring within their interiors and so more properly come under the heading of active galaxies (Section 9.5).

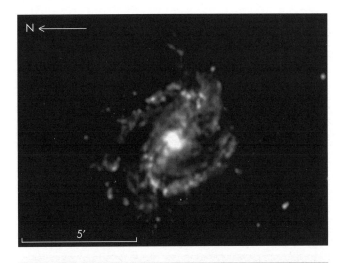

Figure 9.17   A long-exposure CCD image of M83.

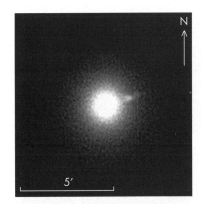

Figure 9.19   A long-exposure CCD image of M87 showing its jet.

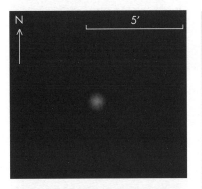

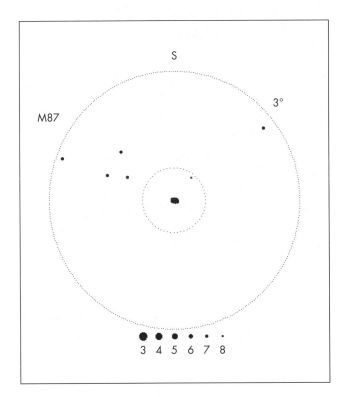

Figure 9.18   Finder chart for M87.

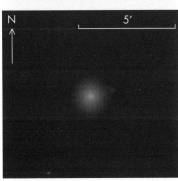

Figure 9.20
The visual appearance of M87 through 3-inch (75 mm), 6-inch (150 mm) and 12-inch (0.3 m) telescopes.

Figure 9.21 M31 and its two satellite galaxies M32 and M110.

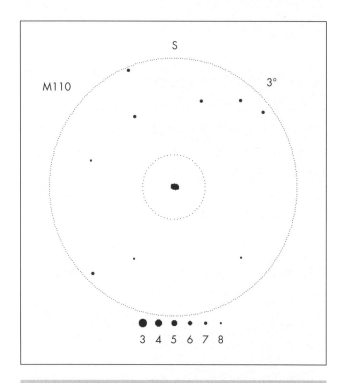

Figure 9.22 Finder chart for M110.

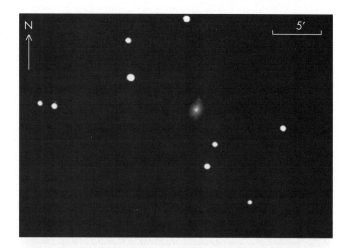

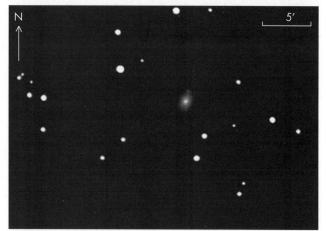

Figure 9.23 The visual appearance of M110 through 3-inch (75 mm), 6-inch (150 mm) and 12-inch (0.3 m) telescopes.

Irregular galaxies tend to be smaller than average and often contain large numbers of gas and dust clouds.

M82 (NGC 3034)   RA$_{2000}$: 09 h 55.8 m   Dec$_{2000}$: +69°41′
Visual Magnitude: 8.4; Size: 5′ × 11′; Distance: 3 Mpc (10×10^6 ly); Type: Irr

With a low-power, wide-angle eyepiece on your telescope both M82 (see Fig. 9.24) and the spiral galaxy M81 in UMa may be seen in the same field of view. The two galaxies are also physically close in space and may have had a near collision some 200 million years ago. M82 is an irregular galaxy that shows signs of activity near its centre. It is also the radio source 3C231. Quite a lot of structure can be seen, though the central dark lane (see Fig. 9.25) is unlikely to be discerned using a 3-inch (75 mm) telescope.

Large Magellanic Cloud (LMC)   RA$_{2000}$: 05 h 24 m   Dec$_{2000}$: −70°
Visual Magnitude: 0.1; Size: 9° × 11°; Distance: 52 kpc (170×10^3 ly); Type: Irr

The Large Magellanic Cloud is the nearest galaxy to the Milky Way. It is easily visible to the naked eye in the

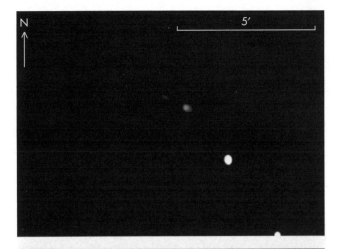

Figure 9.25   The visual appearance of M82 through 3-inch (75 mm), 6-inch (150 mm) and 12-inch (0.3 m) telescopes.

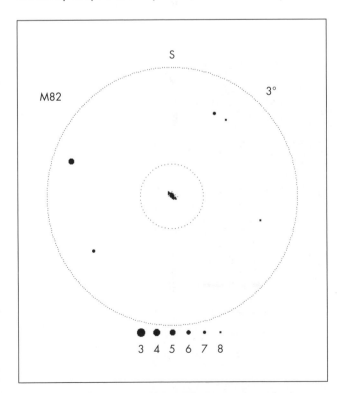

Figure 9.24   Finder chart for M82.

constellation of Dorado. The LMC has about 7 per cent of the mass of the Milky Way, and contains many young stars and much gas and dust. Although generally classed as an irregular galaxy, some observers claim to be able to recognise an incipient barred spiral structure. Along with the smaller and slightly more distant Small Magellanic Cloud (SMC) it has far too large an angular size to be seen through a telescope. The naked eye or binoculars show it to its best (see Fig. 9.26). Many individual nebulae (Chapter 8) and star clusters (Chapter 7) can be seen within it.

9.5 Quasars, Seyfert Galaxies and Other Active Galaxies

This group includes some of the most exciting objects to be found, and their study is one of the leading areas of current research in astronomy. The properties of these objects seem to be related to the hypothesised existence of black holes near their centres. The black holes could have masses up to several hundred million times that of the Sun. Matter falling into the black holes causes intense emission at many wavelengths, and often results in the ejection of material at velocities close to the speed of light in narrow streams called jets. The intense emission from around the black hole causes this region to be far brighter than the rest of the galaxy. With Seyfert galaxies, therefore, there is a small bright centre to the nucleus which dominates the visual appearance. With quasars (quasi-stellar objects or QSOs), the central emission is all that can be seen,[2] and as the name suggests the object has the visual appearance of a star. Many of these objects are bright radio sources and in the radio region often appear as a double-lobed structure far larger than the visual object. This radio structure arises as the high-velocity jets interact with the inter-galactic matter. They are also often bright in the X-ray region, and at all wavelengths they can vary in intensity on time scales ranging from a few days to years.

---

[2] For a few quasars and related objects like BL Lacs, using large telescopes and highly specialised techniques, the underlying galaxy has been detected.

Figure 9.26 The Large Magellanic Cloud (reproduced by courtesy of the European Southern Observatory, ESO Schmidt photograph).

Unfortunately, very little of this activity is visible with a small telescope. The very bright nuclei mean that even though they may be very far away, they can be seen using surprisingly small apertures. However, in most cases, little beyond the star-like centre of the nucleus will be discerned. Nonetheless there is a certain thrill in finding an object like 3C273, and knowing that the photons you are receiving have been travelling through space for over half the lifetime of the Earth – indeed they were 98 per cent of their way here when dinosaurs still roamed the land!

3C273   RA$_{2000}$: 12 h 29.1 m   Dec$_{2000}$: +02° 03′
Visual Magnitude: 12.8; Size: star-like; Distance: about 800 Mpc (2600×10^6 ly); Type: QSO

From a good site on an excellent night it is just about possible to find 3C273 (see Fig. 9.27) in Virgo using a 6-inch (150 mm) telescope. However, a 12-inch (0.3 m) or larger telescope is needed for it to be seen with any ease (see Fig. 9.28). It appears star-like, and you will need a star chart to identify which of the objects in the field of view is 3C273.

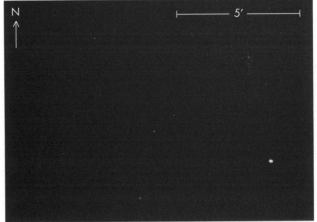

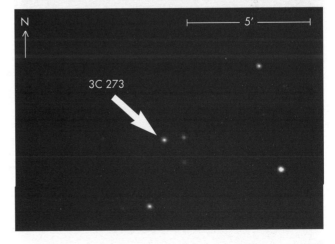

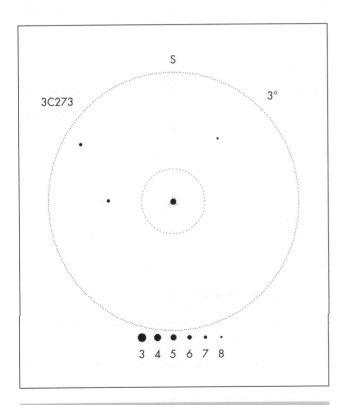

Figure 9.27   Finder chart for 3C273.

Figure 9.28   The visual appearance of 3C273 through 3-inch (75 mm – not visible), 6-inch (150 mm – marginally visible) and 12-inch (0.3 m) telescopes.

M77 (NGC 1068, 3C71) RA$_{2000}$: 02 h 42.7 m Dec$_{2000}$: −00°01′
Visual Magnitude: 8.8; Size: 6′ × 7′; Distance: about 23 Mpc (75 × 10^6 ly); Type: Seyfert

In a very small telescope, M77 in Cetus (see Fig. 9.30) appears almost star-like, since the intense centre of the nucleus is all that may be seen. However some structure may be glimpsed in a 6-inch (150 mm) telescope and the brighter parts of the spiral arms become detectable using a 12-inch (0.3 m) telescope (see Fig. 9.30).

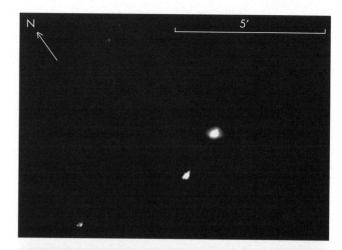

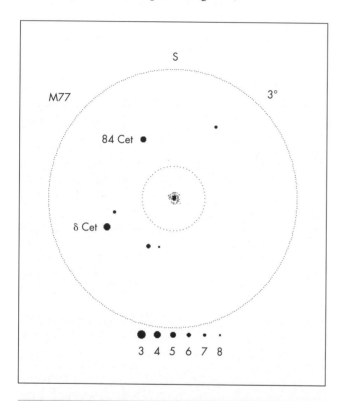

Figure 9.29 Finder chart for M77.

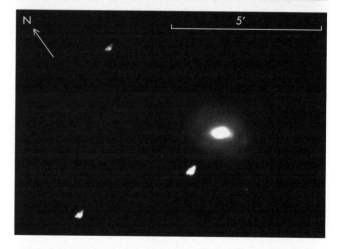

Figure 9.30 The visual appearance of M77 through 3-inch (75 mm), 6-inch (150 mm) and 12-inch (0.3 m) telescopes.

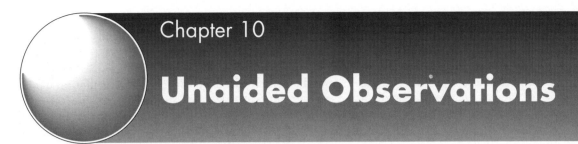

Chapter 10
Unaided Observations

10.1 Introduction

There are numerous observations to be made without the use of telescopes or binoculars; after all, astronomers for 90 per cent of the time that the subject has existed have had to do just that. Some of the phenomena discussed below are perhaps not strictly part of astronomy but will still be of interest to sky watchers. These observations should not be regarded as the poor relations of those made with optical aids, because many of them could not be made using a telescope.

10.2 The Moon

Several phenomena visible to the naked eye, such as the lunar phases, lunar eclipses, the "Man in the Moon", and the "Old Moon in the New Moon's arms" have already been discussed in Chapter 4. The movement of the Moon around the sky, and the possibility of tracking its 18-year cycle have also been mentioned there. An additional spectacular sight on many a clear cold winter's night is the "ring around the Moon". This is a white ring about 22° in radius and centred on the Moon. It is 1° or 2° wide, and can be very bright. It arises from refraction of the moonlight by ice crystals high in the Earth's atmosphere. The solar halo (Section 10.3) is the same phenomenon. The lunar corona arises from scattered moonlight by water droplets in low clouds. It is a variable circle a few degrees across. Both the ring around the Moon and the lunar coronae are indicators of a hazy (or worse) sky. You

might just as well spend your time observing them therefore, because the conditions are going to be too poor for any other type of astronomy!

Another phenomenon associated with the Moon is that of the tides. These are usually regarded as an effect occurring in water. In fact, they occur in the solid surface of the Earth as well. It is the ability of the sea water to flow, and of local geographical structures to amplify that movement, that results in the much greater tidal movement of the water compared with that of the land. Both the Sun and the Moon produce tides, with that from the Moon being about twice as strong as that from the Sun. At new Moon and full Moon, the lunar and solar tides act together and produce the fortnightly spring tides. At either half Moon, the lunar and solar tides are in opposition, and we get the neap tides. If you live near the sea, the tides can form a fascinating study. Although the basic phenomenon arises from the movements of the Sun and Moon, local geography will affect the timing and height greatly, and storms and changes in air pressure all have their influences.

10.3 The Sun

The Sun itself should never be observed directly (see warning in Chapter 3) either with or without optical aids. Various phenomena associated with the Sun, however, may be observed.

A halo may appear around the Sun due to refraction by ice crystals in exactly the same way as the ring around the Moon (Section 10.2). Although this probably occurs as frequently for the Sun as for the Moon, it is much less often seen, because we are normally accustomed not to look close to the Sun in the sky. The greater intensity of sunlight means that other rings and arcs arising from multiple reflections and refractions in the ice crystals can become visible. These include a 46° halo, an arc near the zenith, a circle parallel to the horizon and going through the Sun, anti-Suns and various other complexities, but they are rarely to be seen. A closely associated phenomenon, which is quite common, is the *parhelia* or *sun dogs*. These may most often be seen when the Sun is fairly low in the sky and there is some thin cloud around. They are refracted images of the Sun and they appear at the same altitude as the Sun and at about 25° on either side of it. They can be quite bright, and often appear as a short spectrum. They arise through refraction in columnar ice crystals which have been aligned vertically by air movements.

Much more familiar to most people than these solar halos and arcs, is the rainbow. This arises through refraction in water droplets and is a circle of radius 42° about the anti-Sun point in the sky. The secondary rainbow, which has the colours reversed, is a circle of radius 50°. Below the primary rainbow, supernumerary arcs can be seen at times and these arise through interference effects. In mountains, or from an aircraft, what appears to be a completely circular rainbow can sometimes be seen. The shadow of the observer's head will appear in the centre of the halo. This is not a true rainbow, and its angular size is usually only a few degrees. It is called the *glory*, and it arises through scattering of sunlight by water droplets.

If you are able to watch[1] the last few seconds of the Sun setting over a clear sharp horizon, then you may be lucky enough to see the *green flash*. This is when in the very last second or so of the setting Sun, the upper limb of its disk flashes an emerald green before finally disappearing. It occurs because refraction in the Earth's atmosphere splits the solar image up into multiply coloured images covering the whole spectrum. At the time of the green flash, the red image of the Sun has set, and the blue image is absorbed by the atmosphere, leaving only the green image to be seen.

Finally there is the movement of the Sun in the sky. It is fairly straightforward to construct a simple sundial if you

---

[1] This does, of course, go against all the warnings about not observing the Sun directly. When the Sun is rising or setting, however, it is sufficiently dimmed by atmospheric absorption to be fairly safe to observe. Nonetheless, you may still get an after-image and should not make this a common practice.

have reasonable DIY skills. Instructions on laying out the dials for both horizontal and vertical designs may be found in sources listed in Appendix 2. Of more interest, perhaps, would be the design and construction of a sundial which corrected for the irregular motion of the Sun around the sky and so gave civil time directly. Sundials of this latter type can be quite complex and some of them look more like modern art sculptures than clocks. The irregular motion of the Sun can best be revealed by plotting the analemma. This is a figure-of-eight shape which is obtained by observing the position of the Sun at fixed civil time (alterations to civil time for daylight saving during the summer should be ignored for the purpose of this investigation) throughout a year. It can be plotted out on the ground by marking the top of the shadow of a stake, or on a ceiling by marking the spot reflected from a small mirror placed on a south-facing window sill or by countless other approaches that will suggest themselves to you. Whatever method is chosen, however, it must remain in place for a year. From the analemma you will find that the difference between time on a sundial and civil time, known as the *equation of time*, may be as much as 16 minutes.

10.4 Meteors

Meteors or *shooting stars* are the streaks of light produced as a particle of dust hits the Earth's atmosphere at a velocity ranging from ten to several tens of kilometres per second. The brighter meteors are called *fireballs* and if they appear to explode, *bolides*. Very rarely the particle producing a really bright fireball may partially survive its passage through the atmosphere and reach the surface of the Earth as a meteorite.

Serious meteor observing is a relaxing pastime, since it involves lying down in a deck chair, well rugged-up and with a plentiful supply of hot coffee, and gazing at the night sky. On a clear moonless night about 6 to 10 meteors per hour are likely to be seen distributed all over the sky. These are the sporadic meteors. Their height above the surface of the Earth (which is typically 50–70 km) may be measured by triangulation if two people separated by a few kilometres observe the same meteor. Each observer should note the time of the meteor and plot its track on a star chart. The track can be determined quite accurately by aligning a piece of string along the trail of the meteor and extending the line until it meets bright stars at both ends.

Rather more exciting are the meteors occurring in meteor showers. Meteor showers occur when the Earth passes through a cloud of dust particles left behind by a comet. Then, from 10 to 100 000 meteors per hour may be seen. Furthermore, these meteors will appear to diverge from a point in the sky called the *radiant* (see Fig. 10.1). This is an effect of perspective because the dust particles are all moving on parallel tracks through space. The radiant is the direction of the relative velocity between the Earth and the dust particles. Meteor showers are named for the constellation in which the radiant is to be found. The more prominent meteor showers are listed in Table 10.1. The zenithal hourly rate (ZHR) is the number of meteors likely to be seen if the radiant were at the zenith. Meteors occurring below the horizon and atmospheric absorption mean that these rates will be reduced by a factor of 2 if the altitude of the radiant is 25° and by a factor of 5 if its altitude is 10°. For some showers, the ZHRs are very variable; for the Leonids, for example, rates ranging from 20 to 150 000 have been seen.

10.5 The Milky Way, the Zodiacal Light and Aurorae

These rather disparate phenomena are linked here because observationally they are all faint amorphous glows in the sky.

The *Milky Way* is the part of our own galaxy immediately surrounding the Sun. It is not the whole galaxy because dust clouds limit the distance that we can see within the plane of the galaxy to two or three thousand parsecs. Since the Sun is towards the outer edge of the spiral arms of the galaxy, we see the Milky Way as a faint band of light some 5° to 15° across, stretching right the way around the sky (see Fig. 10.2, *overleaf*). It passes through the constellations Cassiopeia, Cygnus, Sagitta, Aquila, Scutum, Sagittarius, Scorpius, Ara, Centaurus, Vela, Puppis, Monoceros, Taurus, Auriga, Perseus, and back to Cassiopeia. On a clear moonless night it can be seen fairly easily. To observe its details however needs a good site, free of light pollution, and to allow your eyes to become completely dark-adapted. Then it will be possible to see that the Milky Way is quite irregular, with light and dark patches, and split into several branches in places. These irregularities arise from the dark absorbing dust clouds, and from the spiral arms of the galaxy near the Sun having several components.

The *zodiacal light* is sunlight scattered by dust particles which are concentrated within the plane of the solar system. It is therefore aligned with the ecliptic and lies within the zodiac. It appears as a faint cone of light stretching up from the horizon near to where the Sun has set, or to where it will rise. Like the Milky Way it requires a good, light-free site and dark-adaption to see it well. It is possible to trace it right round the ecliptic, but usually only the portion close to the Sun will be seen. The zodiacal light has a slight brightening around the anti-solar point. This arises because direct back-scattering of light by the

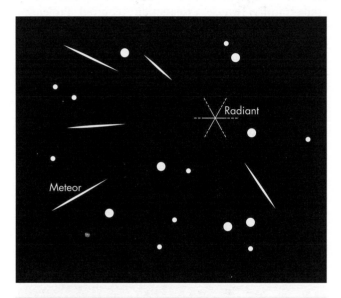

Figure 10.1 The radiant of a meteor shower.

Table 10.1. Some prominent meteor showers

Name	Dates of occurrence	Maximum ZHR
Quadrantids	1–6 January	100
April Lyrids	19–24 April	15
η Aquarids	1–8 May	20
June Lyrids	10–21 June	10
δ Aquarids	15 July–15 August	35
Perseids	25 July–18 August	70
Orionids	16–26 October	30
Taurids	20 October–30 November	12
Leonids	15–19 November	≥ 20
Geminids	7–15 December	60

Figure 10.2 Part of the Milky Way seen soon after sunset.

dust particles is more efficient than that at other angles. This glow is called the *gegenschein*, and is a very difficult object to observe indeed. If you do find it then you can congratulate yourself.

The *aurora* (*aurora borealis* in the northern hemisphere and *aurora australis* in the southern hemisphere) is due to energetic ions and electrons coming from the Sun colliding with atoms high in the Earth's atmosphere. The particles are channelled by the Earth's magnetic field and aurorae therefore mostly occur in regions around the magnetic poles of the Earth. The magnetic poles are at 76°N, 101°W and 66°S, 140°E (Bathurst island in northern Canada, and just off Adelie land in Antarctica), and the peak auroral activity is in zones of about 20° radius around these points. Only observers reasonably close to these zones are likely to see aurorae with any frequency. The aurorae can be quite bright so that the precautions required for observing the Milky Way and zodiacal light are not so important. Individual aurorae come in many forms, with "curtains", glowing patches and "search lights" being typical patterns. They usually move and

change shape rapidly. Auroral activity as a whole varies with the sunspot cycle because the solar flares which produce most of the ions and electrons causing aurorae are most common at sunspot maximum.

10.6 Comets and Planets

Naked-eye observations of the planets are comparatively limited in scope. Their movement across the sky can be plotted over a period of time. This will reveal their retrograde motion whose explanation required Ptolemy's complexity of deferents and epicycles etc. within the geocentric model of the Solar System. Possibly of more interest, certainly more of a challenge, is the naked-eye observation of Mercury and Uranus. Mercury is quite bright but it is always close to the Sun in the sky. Unless you are observing from an excellent site, it is therefore often a difficult object to find. Uranus was, of course, discovered only with the use of a telescope. It is however just bright enough near opposition at +5.5^m to be visible under the best observing conditions.

Comets occasionally are large enough and bright enough to be seen with the naked eye. The development of their head and tail and their movement across the sky are then well worth observing and sketching with as much accuracy as possible.

10.7 Spacecraft

Many spacecraft can be observed with the naked eye. Indeed they are quite difficult to find and follow in a telescope because of their motion (except, of course, when you have just obtained a brilliant photograph or CCD image, then a spacecraft is almost certain to have left a trail across it! – Fig. 10.3). Spacecraft appear star-like but moving quite rapidly across the sky, usually but not always from roughly west to east. Typically a spacecraft would take about 10 minutes to go right across the sky. Predictions of the rising and setting times for a few spacecraft are published in some national newspapers and in astronomy magazines (Appendix 2), but many more than these may be found. High-flying aircraft can be confused with spacecraft, though you will soon get to know the main flight paths in your area if you observe regularly. You can often

Figure 10.3 The trail of a spacecraft and the Whirlpool galaxy (M51).

confirm an object as a spacecraft rather than an aircraft by watching for its brightness to fade as it enters the penumbra of the Earth's shadow, and then disappear altogether as it enters the umbra. Many national astronomy societies have sections devoted to the observation of spacecraft, and if you are interested in this topic it would be a good idea to contact them. Not only will you then get predictions for many more spacecraft, but your observations can be used to monitor the changes in those spacecrafts' orbits.

10.8 UFOs

Once it becomes known that you are interested in astronomy, you are certain to be asked about unidentified flying objects or UFOs. Also as someone with an interest in the sky and spending time looking at it, you are more likely to see a UFO than the average person. In the strict sense of the term, UFOs must clearly exist, since any object moving through the atmosphere which you do not recognise is by definition a UFO. Most such incidents however will turn out to be high-altitude aircraft or balloons, birds, clouds, aurorae etc. The planet Venus is also often reported as a UFO when at its brightest as an evening star. If you think you have seen something which does not have a conventional explanation, then there are societies in most countries to which your observations can be reported. Your chances of having an alien spacecraft land nearby and whisk you off to the far side of the Andromeda galaxy however remain very remote.

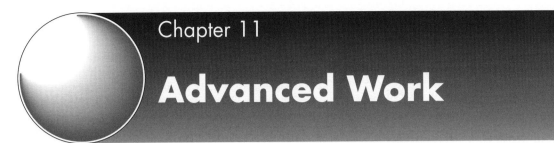

Chapter 11

Advanced Work

11.1 Introduction

While many a pleasant night can be spent tracking down the objects described in the previous chapters just for the joy of detecting them by eye, you will probably soon be ready to try out more advanced observations. Some of these may be carried out using your eyes alone, while others will involve making some sort of objective permanent record, such as taking a photographic or electronic image, or producing a data set of brightness changes against time. If carried out carefully there are many ways in which they can be useful to researchers, and, as has often been noted, astronomy is a field in which amateurs play an active and vital role. This chapter gives advice on extending your observing techniques, either using more specialised equipment or to make particular types of observation. Most of the accessories described are available commercially to suit a range of telescope fitments, or some you could construct yourself. You will also find members of local astronomical societies full of advice, and it would also be a good idea to join the observing sections of national organisations as they will have a range of useful projects to hand.

11.2 Nebular and Light-pollution Filters

11.2.1 Introduction

Ultimately, the visibility of any object depends on it having a sufficiently large contrast against the background sky. Apart from the surface brightness and angular size of the object, this is influenced by factors such as the transparency and tranquillity of the atmosphere, intrinsic skyglow, the presence of light pollution and local stray light. In most situations there is little you can do about these, but fortunately specially designed filters are available that can cut back the sky brightness and thus make the object easier to see. The best design strategy to employ depends on the spectrum of the object and that of the background. If the object has a greater brightness at certain wavelengths then these must be passed through the filter with as little absorption as possible, but parts of the spectrum where the background is brightest should be strongly absorbed as they pass through the filter. In this way the signal from the object is enhanced relative to the noise of the background, leading to an observation of higher quality. Several different types of filter are available tailored for various purposes.

11.2.2 Nebular Filters

By good fortune it happens that the main spectral lines produced by emission nebulae lie at different wavelengths from those produced by background sources, both natural and man-made. Nebular filters exploit this by passing very limited parts of the spectrum, typically including the hydrogen (H)-alpha and beta lines and the oxygen [OIII] lines (Section 8.5.2). Such *narrowband filters* are also good at rejecting man-made wavelengths and are very effective for improving views of nebulae in sites with or without light pollution. It is also possible to obtain *line filters* which select out either the H-beta or the

O[III] emission lines and in some cases H-alpha in addition. If you have a strong interest in observing the faintest parts of nebulae then one of these filters would be invaluable at any site. A filter that includes H-alpha in its passband would also be particularly useful for CCD imaging, as many CCDs are most sensitive in the red part of the spectrum. Your telescope dealer should be able to show you the transmission characteristic of the filter before you purchase, to assist your decision (see also Section 11.3.4).

11.2.3 Light-pollution Filters

Astronomical objects such as stars and galaxies do not concentrate their light emission into limited parts of the spectrum, and instead have a continuous spectrum covering all wavelengths. They benefit best from broadband filters that transmit more of the spectrum than the nebular filters while still blocking streetlights. These filters do, moreover, have good transmission of the nebular emission lines and so act as general-purpose devices giving a significant enhancement of most objects at heavily light-polluted sites (see Fig. 11.1).

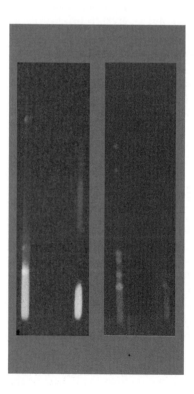

Figure 11.1 The action of a light-pollution filter. Spectra of sources direct (on the left of each pair) and through a filter. The left-hand pair shows the spectrum of a fluorescent lamp, which contains mercury vapour. On the right is a polluted night-sky spectrum showing the same lines, together with emission lines from sodium. The brightest of these is the one that gives the characteristic orange-yellow colour to low-pressure sodium lamps.

11.2.4 Comet Filters

Comets have significant molecular emission bands in their spectra. The interactions between the atoms making up molecules cause these atoms to emit a large number of emission lines. These lie in groups in each of which the individual lines are very close together and seem to form a continuous band. In cometary spectra, the Swan bands due to carbon in the form C_2 are very prominent and filters that preferentially transmit these are available also. Apart from enhancing the visibility of the coma these also allow a comparison of the relative importance of dust and gas in the coma: the dust simply reflects the incident sunlight but the gas emits preferentially in these bands.

11.2.5 Practical Considerations

The filters are available in various sizes and mounting options – some fit directly on to the rear cells of Schmidt–Cassegrains, others are equipped with T-ring threads (Section 11.4.4) and so may be used with any cameras, and smaller ones screw into $1\frac{1}{4}$-inch eyepiece barrels. The type to choose will depend on the design of your own telescope, and some improvisation in mounting the filter may be called for. It is clearly more convenient to have it placed ahead of the eyepiece holder so that it may be used with all your items of equipment. (Trying to swap a screw-in filter from one eyepiece to another in the dark without getting fingermarks on it or dropping it is tricky).

The filters work by exploiting the phenomenon of optical interference and this requires them to be built up of thin layers of material of closely specified thicknesses, which is carefully attended to in their manufacture. Should light pass through them too obliquely, however, the layers appear thicker and the wavelengths of transmission change. This undesirable effect is produced either by tilting the filters or by using beams of light of too great an angle, that is beams from fast telescopes, or from telecompressor lenses. These situations will not stop the filters working, but they will degrade the performance to some extent and should be avoided where possible (see Fig. 11.2). A similar situation will occur if you mount the filters in front of a camera lens for wide-field constellation photography; the best results will be with lenses of longer focal lengths having narrow fields of view. (The aperture can still safely be set to maximum even with a fast lens, as this will not influence the angle at which light passes through the filter.)

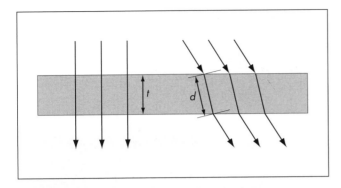

Figure 11.2 Passage of light through an interference filter. Each layer in the filter has a design thickness *t* that is correct for light incident at right angles to its surface, as shown on the left. The light at the right passing through at an oblique angle travels a distance *d* inside the layer and this is greater than *t*. This increases the wavelength at which the filter operates.

For some observations, such as small planetary nebulae which can be difficult to distinguish from stars, it can be useful to "blink" the object by moving a filter in and out of the beam. When the filter is in position the planetary will appear relatively brighter as compared with stars in the view. This can be done by holding the filter between eye and eyepiece, but as there is obviously some potential for scratching the filter, you would be better devising a permanent arrangement if you plan to do a lot of this type of observing. This might take the form of a slide or a wheel to hold the filter, with a second clear aperture position to provide the unfiltered comparison view.

Finally, it must be stressed that attaching a filter always reduces the actual brightness of an object. Coupled with the darker sky background, which makes any stray reflections more annoying, this means that it is crucial to keep light from the surroundings out of your eyes. A simple shield can be made from the lid of a small corrugated cardboard box: it will need to be 6 or 7 cm deep and a bit bigger than your face. (The lid off a box of photocopier paper is ideal.) Make a hole in the centre that is just big enough to pass an eyepiece barrel through and paint the inside matt black or, even better, cover it with black flock paper. Then simply trap it between the eyepiece and the eyepiece holder. It is easy to rotate it to the most convenient position and the telescope controls will still be accessible.

11.3 Colour Filters

11.3.1 Introduction

In contrast to the filters in Section 11.2, these are designed to pass broad ranges of wavelengths and so they appear predominantly of one colour. They are also available in differing degrees of colour saturation. Low-saturation filters have light coloration and pass most of the spectrum to some degree. A pale yellow filter, for instance, passes most light except for some absorption of blue wavelengths. A deep green filter, however, passes hardly any blue or red light.

11.3.2 Types and Fittings

The cheapest filters take the form of thin sheets of coloured gelatine. These have the advantage that they can readily be cut to any size or shape required, so you can adapt them to any situation. They need to be used and stored very carefully, however, as they are easily scratched and are susceptible to damage by moisture. A wide range of colours is available from major photographic dealers.

A more specialised range of coloured glass filters may be obtained from astronomical suppliers. These are mounted in metal rings bearing screw threads enabling them to fit into eyepieces. Looked after with the same care as your eyepieces, these will last indefinitely.

11.3.3 Planetary Observations

Mars and Jupiter exhibit subtly hued features that can benefit from the extra contrast provided by a colour filter. The predominant orange-red surface of Mars, for instance, can be subdued with a green filter to enhance the darker details; the polar caps stand out best with a blue filter. On Jupiter, the belts and spots have such an overall range of colours that a selection of filters has to be used in turn to obtain the best advantage. Venus' cloud features showed up in the ultra-violet in spacecraft images, so it is possible that an extreme violet filter would be of use in trying to detect them. For daytime views of Mercury a red filter will help to suppress the bright blue sky and make the planet's small disk easier to see. A deep red filter is particularly beneficial if a CCD video camera without a short exposure-time setting is used for imaging planets in the daytime, as otherwise it will saturate on the sky.

11.3.4 Imaging

A deep red filter can be useful for improved results when imaging nebulae, as it will pass the H-alpha radiation by which most of their visible light is radiated, but cut out competing background wavelengths. The Wratten #29 filter is particularly useful here; a good photographic shop will be able to obtain one for you. In the case of photography it is important to use a red-sensitive film, such as Kodak Technical Pan (2415) or Ilford SFX, as most black and white films do not respond to such extreme red light. Even the eye is not very sensitive so this filter is not of any great help for visual observations. It is, however, fine for use with CCDs.

11.3.5 Tri-colour Imaging

While colour films can be used for making astronomical pictures, they may not be entirely satisfactory as the colour balance tends to be incorrect unless the exposures are very short (no longer than a few seconds). The best colour photographs are obtained by using panchromatic black-and-white film and taking three separate exposures through red, green and blue filters in turn. Experience with a particular film/filter-set combination will show how to adjust the lengths of the three exposures so as to produce negatives of matching densities. The resulting three negatives are later used to print on to colour printing paper, great care being taken to ensure that the three separate images are registered with one another. As the black-and-white film and also the colour paper will have different responses to the three colour bands, some experimentation will be required to set the colour balance correctly.

The process can be used with CCD cameras as well and is much simpler in this case, as the final image will be produced using a computer. This makes registration of the images very easy and the colour balance can be adjusted an indefinite number of times without using up expensive chemicals and printing paper. One warning must be given, however, and that is that many blue filters also pass deep red and near-infra-red light, and most CCD cameras are very sensitive to this. The "blue" image contains spurious information and will be much brighter than it should be. An extra filter (such as a BG18) is required that blocks out the unwanted long-wavelength part of the spectrum.

Possible targets for tri-colour imaging are gaseous nebulae, where the predominant red colour of emission nebulae will distinguish them from the blue of reflection nebulae, and spiral galaxies where the stars marking out the spiral arms, predominantly blue in colour, are interspersed with red emission nebulae.

11.4 Photography with Your Telescope

11.4.1 Introduction

Astronomers have been using photography to form permanent records of celestial phenomena for almost 150 years and you can do so as well. At the very least, producing a high-quality picture will be a source of personal satisfaction but it may well also contain information of scientific value. An extensive range of specialised accessories is sold to help you. Many complete books have been written on the subject and here there is only space to outline the main points of technique.

11.4.2 Piggyback Photography

A good way to make a start in astrophotography is to mount a camera securely to your telescope and take some pictures of whole constellations. Standard 35–50 mm focal length lenses are perfect for this and any film will do. Adapters can be purchased that fix on telescope tubes to support the camera or you could make your own. (The handscrews that fit into the tripod bush on the camera, can be purchased from photographic shops.) Remember also to affix a suitable counterweight so the telescope remains balanced and inadvertent loosening of a clamp in the dark does not result in a sudden swing of the tube. Often a purpose-built counterweight bar can be bought from the suppliers of your telescope. If you fashion one to your own design, allow for a much greater movement of the weight than is required just by the camera to allow for the addition of extra accessories in the future. Once the camera is fixed, set the focus to infinity and if the aperture (f-stop) is adjustable, set it to maximum size (the *smallest* number).

If the telescope has a motor drive then turn this on, otherwise use the slow-motion controls with a high-power eyepiece to keep a bright star centred in the field of view. The exposure time that you can give will depend on several factors. The best approach is to try a range of exposures – say 5 s, 10 s, 30 s, 1 min, 2 min, 5 min, 10 min – and see which gives the best result. If your camera has a connection for a cable release then it will be easiest to use this; locking cable releases are readily available from camera shops. Otherwise improvisation is called for to hold the shutter down, "Blu-Tack"® or rubber bands work well.

11.4.3 Cameras for Telescopes

Any light-tight box that can hold the photosensitive surface square on to the telescope's optical axis and in the focal plane, and that has a shutter to control the exposure period, will suffice as an astronomical camera. Most people choose, however, to employ a standard 35 mm single lens reflex (SLR) camera in which lenses can be interchanged via a bayonet or screw-thread fitting as this has all the required features, allows easy use of many different types of film and can also be used to take high-quality holiday snaps. When buying you could let your choice be influenced by: (i) being able to find a T-ring (see Section 11.4.4) to match it; (ii) the nature of the focusing screen, plain ground glass possibly with a central clear spot being best; (iii) whether it has a built-in through-the-lens (TTL) light meter; and (iv) whether the reflex viewing mirror can be locked up. It is not likely that the camera will lack a cable release socket but, if possible, fit a cable release to it and check (i) that it screws in snugly, (ii) that it does operate the shutter successfully when depressed and (iii) that if the camera requires the release to be pressed continuously during a time exposure, the lock on the release does indeed hold the shutter open.

11.4.4 Focal Plane Photography through the Telescope

Here the telescope optics are used as a giant telephoto lens, with the film placed at the focus (see Fig. 11.3, *overleaf*). For a focal length of 2 metres the field of view is 40′ × 60′ – ideal for the Moon and larger clusters and nebulae. The first requirement is to fit the camera in place of the eyepiece. The simplest way is to buy a T-ring that matches the lens mount of your camera. One face of this will couple to the camera like a lens while the other carries a standardised ("T") thread. You also need an adapter that fits into the eyepiece holder (or replaces it) and that carries the matching T thread. The overall combination then couples the camera securely to the telescope. With this system you can take your camera and T-ring and fit it to someone else's telescope (or to other optical instruments such as microscopes), and if you should later buy a different make of camera all you need to do is obtain also a new T-ring.

11.4.5 Using Telecompressors

The effective focal length of your telescope may give a perfect scale for photographing some objects but will need modifying for others. The largest nebulae, such as the Orion nebula or the Andromeda galaxy, might require a shorter focal length in order to fit into the film frame, and this can be provided by using a telecompressor lens (see Fig. 11.4, *overleaf*). This is a positive lens that intercepts the beam of light converging to focus at the film and makes it converge more rapidly. This is equivalent to shortening the focal length and is usually designed approximately to halve it. The device can be bought mounted in a cell with T-ring threads on either side so that it fits between the camera and the existing adapter, or there are versions meant for Schmidt–Cassegrains that screw directly to the back of the telescope.

One side-effect of a telecompressor is to make the image on the film brighter, usually by a factor of about 4. This effect is often desirable in its own right, even if the object of interest ends up rather smaller in the picture than you would prefer. It reduces the effect of reciprocity failure (see Section 11.4.8) and is useful for a subject with particularly faint surface brightness, for example the outer regions of a spiral galaxy.

11.4.6 Use of a Tele-extender for Eyepiece Projection

The tele-extender is a tube that is designed to hold the camera firmly at some distance behind the eyepiece. One end carries the thread to match the T-ring and the other either threads (or clamps) on to the eyepiece holder or the eyepiece itself. Sometimes the tube is of adjustable length. The eyepiece is used to project a magnified image on to the film, increasing the effective focal length in proportion to the magnification (see Fig. 11.5, *overleaf*). This is essentially the same procedure on a smaller scale as eyepiece projection for solar work (Chapter 3).

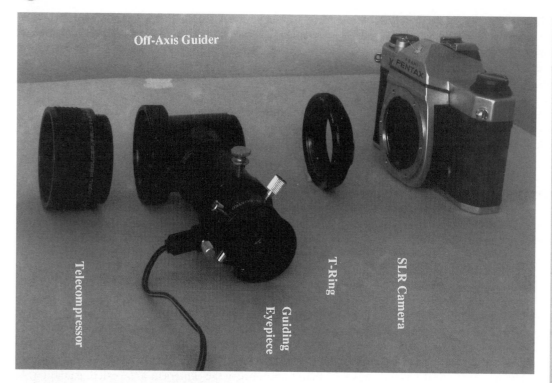

Off-Axis Guider

Telecompressor

Guiding
Eyepiece

T-Ring

SLR Camera

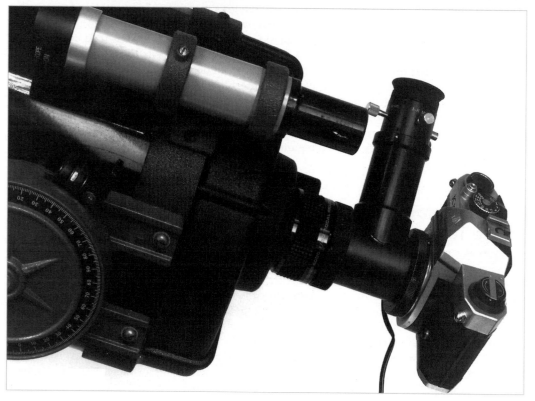

Figure 11.3 The complete set-up for photography.

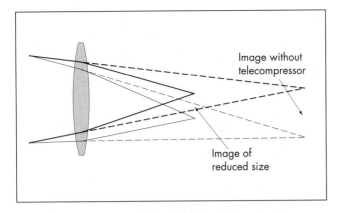

Figure 11.4   The action of a telecompressor lens. The original beam (dashed lines) is made to converge more sharply (full lines), so increasing the focal ratio and decreasing the effective focal length and image size. (The lens is usually of achromatic construction with two lenses cemented together rather than the single lens depicted for simplicity.)

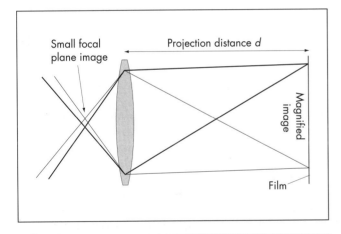

Figure 11.5   Producing a bigger image by eyepiece projection. (The eyepiece is shown as a single lens for simplicity.)

Eyepiece projection is essential for photography of the planets, for close-up views of the Moon and for pictures of close double stars. The image size depends on the focal length of the eyepiece and the projection distance and so it can be varied as required. The bigger the image then of course the fainter it gets, and an upper limit to the magnification is set by the necessary increase in exposure time. Too long and the telescope drive may not be accurate enough or atmospheric turbulence will smear out all the fine detail.

11.4.7 Guiding

If your telescope is of short focal length and has an excellent drive then you may be able to make shortish exposures without any intervention, but it is more likely that you will have to guide. The equipment required for this is discussed in Chapter 2 (Section 2.8.5).

There are different patterns of guiding reticle available, some having a central bull's-eye rather than a complete cross-wire. Only you can choose whether you would prefer to have the star centred in a small circle or silhouetted against a cross. In either case it would be a boon to have the type where the reticle can be shifted around in the field of view, making selection of a guide star easier. Once you are exposing it is simply a case of returning the star to the target as quickly and accurately as possible when it drifts off. You need to be prepared to make the shortest of jabs on the handset buttons.

It is possible to avoid the manual labour of guiding by using an autoguider. This detects the movements of the guide star and operates relays wired into the drive circuits to reposition it correctly. They can use photodiodes or CCDs as detectors. Provided the drive is reasonable to start with and the focus is good, these can produce nice round high-quality star images. A photodiode autoguider often uses a quadrant device having four photodiodes set close together in a cross pattern. The guide star image has to fall over all four detectors and so may need defocusing to make it big enough. The signal from each diode is monitored and the telescope moved until the four signals are equal in magnitude, thus keeping the star image in a constant position on the device.

11.4.8 Choice of Film and Processing

In your first photographic trials it would be a good idea to use a fast colour print film. ISO 1000/31 films have an adequate sensitivity and resolution, and you can take them down to the local photoprocessing shop, remembering to warn them about what to expect. Slide films also work very well and tend to offer greater contrast than print films. If you wish to take pictures for their scientific value, however, you should use black-and-white film and undertake your own developing. Medium-speed films ISO 400/27 are very suitable, particularly if of the modern "T-grain" type. When you need the highest resolution a slower fine-grain film will be best. In cases where the long-wavelength red response is important (gaseous nebulae or solar work with an H-alpha filter) the film of choice is Kodak Technical Pan (2415) and this also has superbly fine grain.

The film speed ratings quoted are always given for everyday use with bright light and unfortunately do not apply at the low light levels of astronomical work. For exposures of more than a few seconds the rating is dramatically and progressively reduced so that longer and longer times are needed to produce a denser negative. This *reciprocity failure* can be readily demonstrated by comparing a 20-minute exposure with one of 5 minutes. The difference between the two will not be anything like as much as you might expect. Fortunately the film can be treated – *hypersensitised* – in a way that increases its low-light-level speed and reduces the reciprocity failure. The standard treatment is to "soak" the film at elevated temperature in a mixture of nitrogen and hydrogen gas for several hours. Kits and canisters of gas are available for you to do this yourself, or if you are disinclined to dealing with potentially hazardous gases, films can be purchased already treated.

Producing a satisfactorily exposed piece of film is only half the job and the processing requires equal care. The necessary tanks, chemicals etc. can be found in a good photography shop and it is not vital to have a darkroom, as it is possible to buy developing tanks that can be loaded in (subdued) daylight.

11.4.9 Projects

The range of projects is limited only by your imagination and those suggested here are just starting points. Most of them apply to CCDs as well.

> Your own complete set of those constellations visible from your latitude.
> All the Messier objects visible from your site, or all those of a particular type.

> Views of all the planets.
> A set of views showing the changing phases of Venus.
> An atlas of lunar highlights.
> A lunar feature under many different angles of illumination.
> Measuring the rotation periods of minor planets by changes in brightness.
> As complete a record as possible of a cometary apparition.
> Finding the times of minima or maxima of variable stars.
> A set of galaxies of differing Hubble types.
> Taking part in supernova patrols.

11.5 CCDs

11.5.1 Introduction

Just as photography revolutionised astronomy in the nineteenth century, so the charge-coupled device, or CCD for short, has revolutionised it in the late twentieth century. More remarkably, the power of the CCD has become available to amateur astronomers at a cost similar to that of a quality telescope. This has enabled increasing numbers of amateurs to make observations to professional standards and to contribute to scientific research. Compared with photography, CCDs have several significant advantages: a greater efficiency in the detection of light, a linear response to light, a greater range of brightness which can be recorded faithfully, a better dimensional stability, easier extraction of data from the image, immediate viewing of an image as soon as it has been taken, and no need for any manipulations in total darkness. The main disadvantages are the initial outlay to buy the camera and computer and the relatively small size of present devices compared with the area of a 35 mm film frame. The general equipment set-ups discussed in Sections 11.4.4 to 11.4.7 apply also when using CCDs. Most CCD cameras come with a coupler that allows them to fit in place of the eyepiece, and this means they are held in with the normal small fixing handscrew. For security an adapter that couples to a T-ring just like an SLR camera can be purchased, and this has the additional advantage of making the cameras interchangeable. You could then use ordinary camera lenses with the CCD for wider field views. An additional difference is that CCD cameras do not come equipped with reflex viewers and this, coupled with the small chip size, can make it difficult to get objects into the viewer. This problem may be overcome by buying or making a special viewer with a flip mirror to fit between the telescope and camera. The viewer could also incorporate filters for tricolour work and a telecompressor lens to increase the field of view.

11.5.2 Camcorders

Many people already possess a CCD buried inside a camcorder and it is worth experimenting with this. The Moon and bright planets make ideal subjects. First focus the telescope by eye, then set the camera focus at its greatest distance (turn off its automatic focus), and hold it up to the eyepiece. The view can be recorded for later viewing and could also be fed into a computer video capture card for digitisation and subsequent analysis. One particular advantage of this technique is that it takes a very large number of separate images and some of these will have been taken at moments when the atmosphere was steady. You can sort these out from the rest at leisure by examining the tape frame-by-frame.

11.5.3 Purpose-designed CCD Cameras

While the short exposure (max. 1/25 s) of a camcorder chip works fine for bright objects, much superior results will be obtained with a camera designed expressly for astronomical use. This should have several important features: a means of taking a time exposure (of perhaps up to an hour's length); a means of cooling the CCD chip, preferably to −40°C or lower; a high-quality digitiser to convert accurately the image to digital form for display on a computer; a computer program to simplify the operation of the camera; and a means of storing the numbers representing the image permanently on to computer disk.

Although the CCD chips in cameras all work in similar ways, the overall system varies from camera to camera. Some are intimately connected to your computer, carrying some of their circuits on an expansion board. Others include complete computer systems that communicate with your main computer via a data link. These can be convenient for use with laptop computers that do not have room for interface boards. Also there can be a greater distance between the camera and computer, which

is handy if you want to remain indoors while the telescope operates out in the cold. The drawback is the time needed for the transmission of the image information (which can be particularly irritating when trying to focus the camera).

11.5.4 CCD Chips

The CCD chip itself consists of a large number of pixels – individual light-sensitive areas – set in a regular array. Typically the array would be a few hundred pixels on each side with each pixel in the region of 10–20 μm across. This makes the overall size typically about 10 mm across the diagonal, which corresponds to a corner-to-corner field of 0.3° if used with a focal length of 2 m. Each pixel is delimited by tiny electrodes which are connected to circuits built around the edge of the chip that energise them in a variety of ways (see Fig. 11.6, *overleaf*).

11.5.5 Sensitivity to Light

A CCD camera has a higher efficiency for the detection of light than photographic emulsions so its effective speed rating is quite high compared with most films. This varies according to the size of the pixels, the chip temperature and the quality of the charge-measuring circuit. Larger pixels each collect more light and overall this results in a better speed rating, particularly for extended objects. Cameras are often designed in such a way that you can ask the software to lump the normal pixels together, thus making it super-sensitive. Of course the resolution is badly degraded but it is a useful technique for acquiring and centring an object. Exposures of a few seconds easily show the Messier objects and, as the number of data points is much smaller, the processing and data transmission times are reduced.

11.5.6 Astrometric Measurements

The pixel array of a CCD produces a precise set of reference points that make it easy to measure the relative positions of objects in the field. You could find the position of a comet or asteroid, for instance, by comparison with background stars. Since the publication of the *Hubble Guide Star Catalogue* it has become easy to find reference

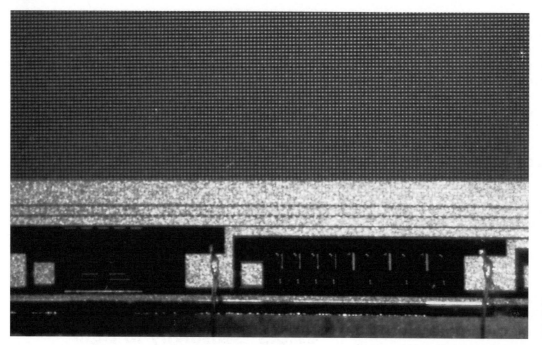

Figure 11.6 Part of a CCD chip. The upper section contains the pixel array, and below this some of the supporting electronics and connections can be seen. The pixels of this chip are 22 micrometres square.

stars with accurately known positions even in the small areas covered by CCD images. Comparison of these with the images will allow the determination of the image scale in pixels per arc second and of the orientation of the field (i.e. the direction of North in the image). From this information the object position is found. Similarly the spacing and position angle of suitably wide double stars could be ascertained.

11.6 Photometry

11.6.1 Introduction

One of the areas that has traditionally been strong for amateur astronomers is the estimation of the visual brightnesses of variable stars (Chapter 7). There are so many of these that it is impossible for researchers to keep tabs on them. While some behave in more-or-less predictable ways, many are unpredictable and ideally need to be observed by someone at least every few years. With some practice the brightnesses can be estimated to 0.1 magnitude with no more equipment than the eye and a prepared chart. Most people worked this way in the past because the alternative of objective measurements involved either photography with its continual running

costs and time-consuming processing, or using photomultiplier tubes. The latter are damaged by bright light and required bulky very high-voltage power supplies with obvious hazard. Nowadays photodiodes are available that only need low voltages, and advances in electronics mean that even photomultiplier tubes can be reliably protected against excessive light and can have power supplies and connections completely enclosed within the photometer. Furthermore, the wider use of computers simplifies the collection and analysis of data.

11.6.2 Photographic Photometry

Photography has running costs, requires an investment of time outside clear sky periods for processing and examination of the negatives, and without specialised instruments does not seem to offer a significantly improved accuracy compared with visual estimates. As a result its use for photometry is not widespread outside professional observatories, who possess microdensitometers to measure the degree of blackening of the negatives. Without these the film must be examined with a hand lens to intercompare the star images, thus duplicating the labour that would be necessary if doing the work directly by eye at the telescope. Of course, being able to do this in comfort and without danger of interruption by the weather should improve the quality of the measurements. Also it will be possible to work on fainter stars, though to get significantly beyond the visual limit for the instrument requires pin-point guiding. Unhappily, observers with bright skies will find themselves stymied by the background fogging of the negatives, making it more difficult to locate the boundaries of the star images.

One area in which photography has been useful is the determination of the times of minima of eclipsing binary stars. Observations over a few hours might be needed and it is asking a lot of an observer to maintain objective accuracy over such a period. Film has no such worry, and even the bother of remembering when to take the next exposure can be eliminated by fairly simple automation of the camera. Several images of the variable and its neighbours can be taken on one frame; separation is achieved either by interrupting the camera drive for a short while between exposures, or by deliberately mis-aligning its polar axis, thus introducing a steady drift.

11.6.3 Photometers

Whatever it uses as its light-measuring element, a photometer (see Fig. 11.7) needs to have certain features: (i) a means of restricting the area of sky whose light is admitted to the detector; (ii) a means of viewing this area of sky for setting up; (iii) a (set of) filter(s) to define the colour of light being measured; and (iv) an output reading. This latter can be an analogue or digital meter on the photometer, or a voltage that can be measured using a chart recorder or A/D converter on a computer, or a pulse train that is fed to a counter or frequency meter.

Photometers incorporating these features and suitable for small telescopes can be purchased ready-made. Some models offer cooled detectors for better noise performance and some allow complete computer control.

11.6.4 Filter Sets

Stars of differing physical properties radiate varying amounts of light at different wavelengths and one of the aims of photometry is to utilise this to categorise stars. To achieve this it is necessary to make observations over agreed ranges of wavelength, and it is the job of filters to let through the correct colours of light to the detector. They can be mounted in a slider or wheel, the latter being particularly convenient for automatic control. It is easier if they are mounted behind the position of the viewing optics, as then the eye receives white light which makes

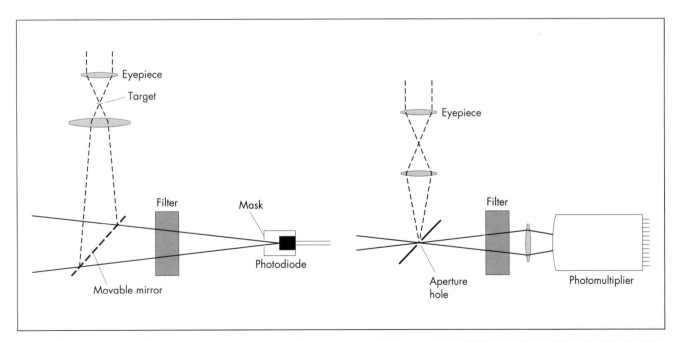

Figure 11.7 Schematic designs for photometers.

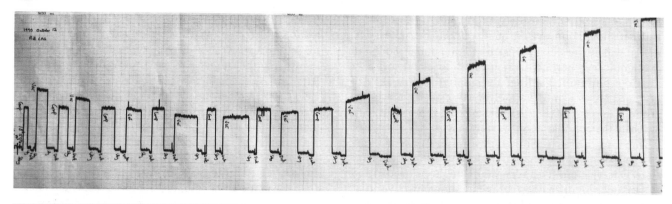

├────── 30 MINUTES ──────┤

Figure 11.8   A chart recording of an observation of the eclipsing binary star RZ Cas. The atmospheric conditions were very stable and the variation is shown well even in this raw trace.

for more comfortable observation. From the point of view of cost they are better in a position close to a focus, where the telescope beam is small and so smaller diameter filters can be used. (They should, however, be in a position where the beam is still several mm across, otherwise any dust particles on the filter surfaces could badly influence the measurements.)

11.6.5 CCD Photometry

Photometry can also be carried out accurately using a CCD. The sequence of measurements needed to compare the variable star with a reference standard star (Chapter 7) takes some time with a conventional photometer. During this time both the amount of light transmitted through the atmosphere and the background illumination of the sky may change. Using a CCD, however, you are able to observe variable, comparison and sky all at once and so any changes affect all measurements by pretty much the same amount. The actual photometry is carried out after acquisition of the image, by using a sort of "virtual photometer". An area of the image can be measured by moving around a box analogous to the stop of a photometer using the mouse or cursor keys. The sum of the pixels in the box gives the total light that "entered the stop". The sky is found by moving the box to a blank area to the side of the star. Alternatively two concentric boxes can be moved together. The inner box covers the star image plus sky, the outer ring between the two boxes just contains

sky so from these two the starlight can be calculated automatically (see Fig. 11.9).

11.6.6 Observing Projects

There are about 2500 known variable stars with maximum of magnitude 9 or brighter, and many of these vary over time scales that are either irregular, semi-regular or slowly changing. They thus need observing at intervals in order for their behaviour to be quantified. Eclipsing binaries are a case where a useful precision is easily achievable, easily exceeding the errors to be found in the ephemerides in many cases. Data to be used in calculating the times of eclipses can be found in an annual publication produced by the Cracow Observatory, Poland (Appendix 2).

Similarly the times of maxima of RR Lyrae variables are worth checking, and the same publication carries predictions for these. If you want to test out your system and be sure of seeing a variation then try CY Aqr, which is a pulsating variable with a period of only 88 minutes. It is not particularly bright (max. 10.4, min. 11.1), so it makes a useful test object (see Fig. 11.9). It is situated near the celestial equator, at $RA_{2000} = 22$ h 37 m 47 s, $Dec_{2000} = +01°32'12''$, and so is observable from both hemispheres.

Occultations of various sorts are covered in Section 11.7.

Many other ideas for observing can be found from organisations such as IAPPP (International Amateur–Professional Photoelectric Photometry).

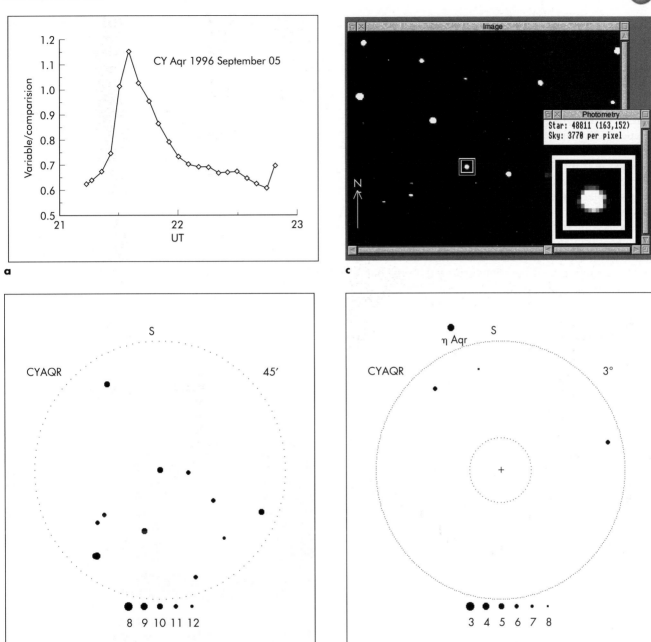

a

b

c

d

Figure 11.9 The rapid variation of CY Aqr derived from a set of CCD images, and finder charts. In part (c) the brightness of a comparison star is being measured.

11.7 Occultations

11.7.1 Introduction

There are so many bodies moving in their various orbits that it is inevitable that occasionally three of them happen to line up. When one of the end ones is the Earth then to an observer at the right place the middle object will pass in front of the third body and blot it out – *occult* it – partially or completely (Chapter 5). Exactly what "in the right place" means depends on the relative sizes, distances and positioning of the objects. An asteroid 50 km across passing in front of a star, for instance, will cast a shadow 50 km wide and anywhere the shadow falls the star will disappear. If it falls across high latitudes, the north-to-south size of the zone measured on the curved surface of the Earth could be quite a bit bigger than 50 km. The shadow will sweep across the surface rather rapidly and the occultation lasts only a few seconds, so you also have to be observing at the right time. If the Moon occults a star the event is seen over a considerable part of the Earth because of the large size of the Moon, and can last up to an hour. The only interesting parts, however, are the moments of disappearance (immersion) and reappearance (emersion), so it is just as important to be looking at the right time! Occultation observations find application in a surprising number of fields: (i) the shape of the Moon and asteroids; (ii) improving the ephemerides and orbital theory of Solar System objects; (iii) probing atmospheres of other bodies; (iv) providing data on multiple stars; (v) measuring the angular diameters of stars; (vi) detecting comets in the Oort cloud; and (vii) analysing the properties of distant galaxies via gravitational lensing. And we should not forget that most spectacular occultation of all, a total eclipse of the Sun, providing precious minutes in which to study the solar corona.

11.7.2 Lunar Occultations

In spite of the number of lunar orbiting craft that have taken images, knowledge of the surface of the Moon is being added to all the time by observations of lunar occultations. Additionally, the positional information that they give is vital in measuring the slow-down rate in the rotational and revolutional motions of the Earth/Moon system which govern international civil timekeeping.

Predicted times of occultations of stars and planets can be found in publications such as the *BAA Handbook*. Times are given for only a few places but instructions are included to find the time more accurately for your own latitude and longitude. The ease of seeing the event varies with the phase of the Moon. The closer the Moon is to full, the more scattered light there is and the illuminated surface gets closer to the limb where immersion will take place. Before full, emersion will occur at the bright limb, making it more difficult to detect, quite apart from the problem of successfully identifying the point where the star will reappear. Clearly immersion offers a problem after full Moon while emersion still requires finding the correct reappearance point (see Fig. 11.10). Taken together with the generally unsocial hour of events during the waning phase, this leads to a decidedly uneven coverage during each month. Anyone prone to insomnia could help out here!

If you decide to try making an accurate timing of an event then you need to obtain time information to at least 1/10 s accuracy. This can be obtained from a short-wave radio or in some places from the Speaking Clock (provided you are not put through to it via satellite). Radio-controlled timepieces are now obtainable at modest cost and these offer high accuracy if kept out of the way of interfering equipment such as televisions and computer

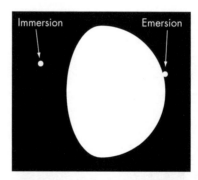

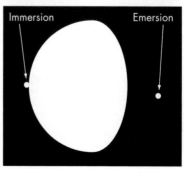

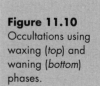

Figure 11.10
Occultations using waxing (*top*) and waning (*bottom*) phases.

displays. You can measure the time of the event using a stopwatch to find the elapsed time from a known time signal, or by recording the time information together with a marker signal operated by a hand button on to a tape recorder or chart recorder.

Photometric observations are also possible and again the time should be accurately recorded. The signals could be registered on a chart recorder or, for easier analysis, fed into an A/D port of a computer. An example of such a trace is shown in Fig. 11.11. This illustrates some of the characteristic features: scintillation of the star together with a steady increase in signal level as the Moon approaches the star, an exponential drop due to the time constant of the photometer amplifier after the star disappears and further steady increase in the signal (with reduced scintillation as the photometer is now viewing an extended area of the Moon's surface rather than the stellar point source). If you have a separate guide telescope then you could previously offset this on to a star that is not going to be occulted and use this to keep the telescope pointed while the immersed star remains behind the

Moon. Then you have the chance to get an observation of its emersion.

Such a trace could also observe the separate steps that occur with close double stars. As the average speed of the Moon is about 0.5 arc seconds per second then a time resolution of 1/20 s should reveal stars separated by 0.05″. (Note that this refers to the projected separation in the direction of the Moon's motion which is not necessarily the actual separation.)

Much precise information is obtained from observations of grazing occultations. There may be two narrow tracks (typically 3 km wide) where the star is barely occulted, just grazing the northern or southern limb of the Moon. In this circumstance the irregular profile of the limb can cause the star to flash off and on as the light is blocked by lunar mountains or shines through valleys. The pattern of flashes is very sensitive both to the surface topography and to the exact position of the Moon relative to the star and observer.

Lunar occultations of planets are more leisurely if less frequent affairs, and offer scope for interesting photographs or images.

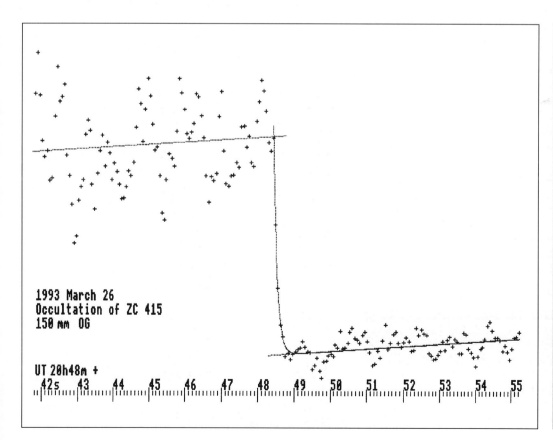

Figure 11.11 A lunar occultation recorded using a computer.

11.7.3 Asteroid Occultations

This is a field in which valuable contributions can be made, either by visual timings or photoelectric recording. The time and place of an occultation give a precise position for the asteroid, helping to improve knowledge of its orbit, while the durations from a number of observers put together show the shape of its shadow, mirroring its actual shape. Predictions of the occultations of suitably bright stars are prepared in advance and the general area where the event should be visible determined. This is usually refined as the event draws closer by precise positional observations. You can find lists of expected occultations for your part of the globe on the Internet, or join the International Occultation Timing Association (IOTA). If you have a CCD you can also help with the astrometric observations needed as an event draws near.

11.7.4 Satellite Eclipses and Mutual Events

The satellites of Jupiter offer interesting observing projects (Chapter 5). Near quadrature, the shadow of the planet lies in such a position that a Galilean moon can be eclipsed in the shadow well before it reaches the limb of the planet. Observations may be made to determine the time of mid-light of these eclipses. Predictions are readily available in the popular magazines.

The systems of both Jupiter and Saturn offer more subtle events at intervals of ~6 and 15 years respectively. At times their orbital planes point towards the Earth, so that the satellites pass in front of one another. The precise timings of these occultations (which may be partial or total) provide important dynamical information about the satellites, improving our predictions of their positions. This is particularly important for the planning of spacecraft missions where close passage by the moons may be required for highest quality imaging.

11.8 Computers in Astronomy

11.8.1 Introduction

Thumb through any recent astronomy magazine and you cannot help noticing adverts for computer software of all sorts. Astronomers have been using computers since they were invented and the modern personal computer can offer you a range of facilities to make your observing life happier. These include the collection of information before you start work and the reduction of data or an image after a successful observation. Of course the responsibility still ultimately lies with you to ensure that you understand what the computer is doing to any data, but programs obtained from reputable sources should be satisfactory. As mentioned elsewhere, the observing sections of national astronomical organisations can help in the procurement of suitable products.

If you use the computer to obtain observations do remember to make back-up copies of the files immediately – you can easily erase them if it turns out that the information is not useful after all. Also try to devise a tidy directory structure – maybe keeping all the observations for one day in a folder whose name indicates the date, such as 97AUG29.

11.8.2 Choice of Computer

If you have already owned a computer for some years then use that for at least a while before deciding if you need to buy another. Shareware is copiously available at small cost and this will give you a feel for whether you are likely to be frustrated by the time taken for a slower computer to perform its operations. If buying one for the purpose, then a second-hand or surplus machine may prove fine if you are not interested in a lot of time-consuming calculations such as minor planet ephemerides or if you do not need an Internet connection. Otherwise it is a good idea to purchase the best specification computer that you can afford and this is now certain to include Internet and multimedia facilities. A laptop is worth serious consideration, as its portability means you can easily take it out of the typically rather insecure environment of the telescope shelter at the end of work. Also it will take up less valuable space on the observing table. A monochrome display will do while observing so you could purchase a normal colour monitor for other uses, which would work out cheaper than buying a colour notebook.

11.8.3 "Planetarium" Programs

This is the name for software that gives you a representation of the sky for a particular time and date on the computer screen. It is similar in effect to the manual planisphere but with the advantages of displaying the positions of the planets, being adjustable for any position on the Earth and allowing the effects of precession to be included so that you can see the appearance of the sky at remote epochs. Some programs also allow you to enter the orbital elements of comets etc. so that they can be plotted as well. Many have databases built in so that clicking over a star will bring up basic information, such as its brightness, coordinates, spectral type etc., and often there are close-up images of planets and nebulae available. These programs are useful for educational purposes and are valuable in planning a night's observing as you can easily predict the altitude of any object as the night progresses. They will also allow you to anticipate close planetary approaches to the Moon or to each another, and the best dates for observing Mercury.

11.8.4 Databases

Many standard astronomical catalogues are available in database form with software to allow easy retrieval of information. This will always include the position of the objects (usually in epoch 2000.0) together with other pertinent details such as brightness, size etc. You can use such data to select objects worth observing with your set-up, and this will extend the number of targets very considerably, when compared with the lists of well-known objects.

It is worth considering using a database as your observing log book, or at least as a fair copy. By setting up a suitable structure you could then immediately retrieve a list of all your observations of, say, Jupiter during any period of time. Another benefit could be improved legibility, particularly if you have had to take notes in a hurry. You may well be able to decipher them a few hours later, but a few years later?

Finally, the contents indexes of both popular magazines and research journals are available in database form and these will help you search for information on specific topics.

11.8.5 Ephemerides

Programs to calculate the position of Solar System bodies for any time and any place are readily obtainable, and on the whole you get what you pay for. The professional program to replace the paper version of the *Astronomical Almanac* will provide high-precision results but is unlikely to be worth the outlay for most people. There are several

programs that are dedicated to minor planets, and these have the orbital data for thousands of asteroids "built in" so that you merely have to enter the name or number of the required object. They will also search for asteroids near particular positions and some are designed to be used as predictors of occultations.

Unless you do want to observe asteroid occultations, cheap and cheerful shareware is probably adequate, even if the user interface leaves something to be desired. There may be small residuals in the results but you are not likely to be caused any trouble by these. It is more important to check for viruses! Obviously you will need to find orbital elements and input them but these are readily obtainable either from the Web (see Section 11.8.8) or in magazines.

If you keep your computer clock accurate, ephemerides will also provide you with the current local sidereal time, though this is usually also given by the planetarium-type programs.

11.8.6 Images and Image Processing

As CCDs have become more widespread, the number of image processing packages around has mushroomed and they incorporate increasingly sophisticated techniques. They all offer certain basic operations such as contrast enhancement (Fig. 11.2, *overleaf*) and the best way to choose one is by convenience of use. You need to ensure that it will recognise the particular file format(s) of your images, and the range of output types for the result may also be important if you wish to share with others. The FITS format is recognised within the astronomical community but is not common elsewhere. It may be used to transfer raw or processed image data. Other formats such as TIFF and GIF are in widespread use and are fine for a final image. As they often limit the range of grey levels that may be present, however, they may cause a loss of precision if used to transmit raw data, so should be used with care for this purpose. Another format that is "lossy" and should be confined to saving final processed images is JPEG. This has the advantage that it offers considerable compression, thus saving on storage space and transmission time. Despite this compression it is capable of producing a displayed image in full colour that is virtually indistinguishable from the uncompressed version.

One area in which computers help considerably is the comparison of images produced at different times, traditionally done by a mechanical blink comparator. This allows the detection of any changes, caused either by

Figure 11.12 The effect of image processing. On the *left* is shown an image as produced by a CCD camera. On the *right* is exactly the same image after some simple image processing.

objects that have moved (comets, minor planets) or by those that have changed brightness (variable stars, novae, supernovae). It is much easier to register the images using a computer to perform any necessary scaling and/or rotation, and additionally scaling in brightness and contrast can be done. The originals need not be from a CCD but could be photographs digitised via a video camera or scanner. Presenting the images sequentially on the screen is no problem, or one could be subtracted from the other to leave strong signals only at the differences. In some cases simple superposition is sufficient to yield interesting results (see Fig. 11.13).

11.8.7 Data Processing

Serious observing programs can result in the need to reduce quite a large amount of data and suitable software will reduce the labour considerably. One particular area that has benefited is astrometric work on asteroids, where a program can find the best-fit positions of several star images together with the target and calculate its position, frequently to better than one second of arc. The programs rely on the data in the *Hubble Guide Star Catalogue*, so you do need a computer with a CD-ROM drive.

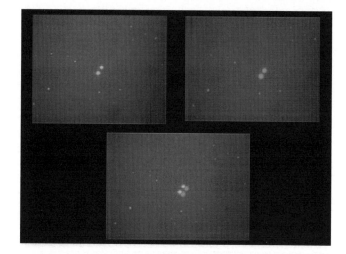

Figure 11.13 Two photographs of 61 Cygni taken 3 years apart. After conversion into digital form and superimposition of the background stars, the large proper motion of this pair is readily apparent.

Photometry using CCD images or photometers has already been discussed. Useful further reduction can be carried out using graphical and curve-fitting packages, to find the best-fit time of minimum of an eclipsing binary for example.

11.8.8 The Internet

One area of computers to which astronomers of all sorts have taken like a duck to water is communication via the Internet. The amount of information available is vast, so much so that you will need to rely heavily on the "search engines" to sift out just that in which you are interested. Here you will find whole textbooks on astronomy, databases, news on new discoveries, images galore, technical information on instruments on the ground and in space, pages of answers to "frequently asked questions" (FAQs), useful software and newsgroups. The latter are particularly useful if you encounter some problem with your telescope or instrumentation, as you can post your problem and in nearly every case someone will reply with useful comments. Alternatively, you can browse the past communications at leisure to see if the question has been dealt with previously. You may prefer this to actually subscribing to a group, as this could generate several mails a day.

11.9 Spectroscopy

11.9.1 Introduction

Spectroscopy is the detailed analysis of the light received from an object, recording the intensity of light wavelength by wavelength. It could be seen as an extension of photometry in which a large number of filters are used in turn, each having a very narrow bandpass. In practice, it is much more than this as the whole spectrum is recorded in one observation. While imaging tells you the overall shape and form of an object (as projected from its 3D shape on to the 2D of the sky), spectroscopy can reveal the chemical compositions, densities, temperatures and motions of its constituents. The spatial distribution of the motions may also allow some inference to be made of the 3D structure. It will be clear from this that observing spectra is a fundamental part of the study of any astronomical body and all major observatories include equipment for such studies. It is therefore unfortunate that this is one field in which the contribution by non-professional astronomers is rather limited. Two of the reasons for this are that accurate spectrographs need to be of massive construction and

require temperature stabilisation, neither of which fits in easily with the average amateur's set-up. The main hope for setting up a good spectrographic capability might lie in utilising a siderostat. Here a flat mirror is used to direct a beam of light towards a fixed telescope which can be mounted together with the spectrograph on a strong table in a shelter. The telescope can still be used for other work, though at the expense of field rotation. Also the equatorial mounting of the flat would need an accurate display of its position to enable you to point it in the correct direction. It can be seen that such a set-up would prove quite costly – the flat mirror needs to be of high quality and must be bigger than the telescope aperture – but maybe the compensation of observing under cover would help to make it worthwhile!

Fortunately observations for your own interest can be made with simpler equipment, and as with all astronomical work there is always the chance of making a valuable serendipitous contribution.

11.9.2 Objective Prisms

The objective prism works in the simplest possible way: it is placed in front of the telescope or a camera lens and starlight enters the prism directly. Thus the input light is parallel and the dispersed output light is focused by the optics as normal. The length of the spectrum produced depends on the type of glass, the shape of the prism and the focal length, but for a standard 50 mm camera lens and an equiangular (60°) prism would only be about 2 mm. Thus a longer focal length would be needed to see any detail. The prism deviates the light by about 40° so that the apparent pointing direction of the camera would have to be offset from the object of interest by that amount. The prism would be mounted with the exit face 20° to the camera, and if smaller than the aperture of the lens, would need a mask to prevent direct light from straight ahead entering the camera. On experimentation a drawback of this simple method will be apparent, namely that the spectra of adjacent stars are liable to overlap. The range of objects that you could observe would be considerably enhanced with a prism large enough to fit over a telescope: apart from the greater light grasp the increased focal length spreads out the spectrum more so detail is easier to see. As so often there is a snag and that is the cost of having a prism manufactured, as "off-the-shelf" items cannot be found large enough, or with the correct shape (the angle between the input and output faces should be only a few degrees).

11.9.3 The Direct Vision Spectrograph

For purely visual observations, an alternative is the direct vision spectrograph. This consists of three or five small prisms cemented together, the glasses being chosen so that the overall path of light is undeviated but there is still dispersion. Combined with an eye lens, this device fits in place of the normal eyepiece of a telescope.

11.9.4 Slit Spectrographs

If you really want to get into high-quality spectroscopy then the slit spectrograph is essential. This overcomes the objective prism problem of adjacent objects interfering by using an entrance slit at the focal plane of your telescope; this slit only lets in the light from one star at a time. Behind the slit the diverging beam is converted to parallel light by a collimating mirror or lens. This parallel beam is then suitable for use with a prism, or more commonly with a diffraction grating. After dispersion another lens refocuses the light into the spectrum. A camera lens is often used for this, so that an ordinary photographic camera can be used for recording the result.

It is possible to buy complete slit spectrographs that are designed to fit on to the larger-sized amateur telescopes. Building your own is possible, but perhaps somewhat daunting in view of the need for mechanical stability, providing an entrance slit probably not more than 1/10 mm wide and a need to arrange optics to view a star on the slit in addition to those for the production of the spectrum.

11.9.5 Observing Projects in Spectroscopy

As noted above, it is instructive to at least observe some spectra and some suggestions are given here.

1. Compare stars of different stellar types. Vega or Sirius, for example, clearly show the hydrogen absorption lines characteristic of early type stars, but these are weak in Capella, whose spectral type is similar to the Sun; Betelgeuse or Aldebaran start to show the "bands" which are the signature of cool late type stars.

2. For a spectacular exhibition of bands, try to observe the long-period variable Mira (Omicron Ceti) at maximum.

3. Emission line spectra are very apparent from sufficiently bright novae; H-alpha, for instance, at 656 nm is difficult to detect visually because the eye is not very sensitive in the deep red, but it can be very bright in a nova spectrum (see Fig. 11.14).

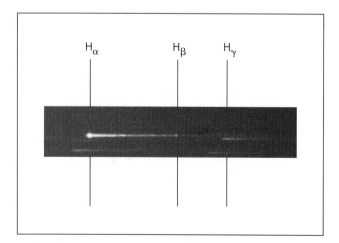

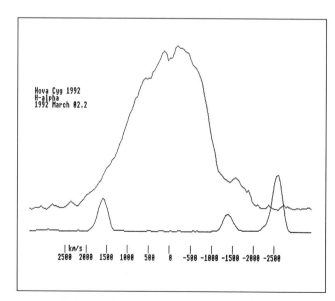

Figure 11.14   The spectrum of Nova Cyg 1992 taken with an objective prism (*left*). At *right* is shown the profile of the H-alpha line derived from an observation made with a slit spectrograph.

4. Compare the spectra of nebulae such as the Orion nebula (M42), the Ring nebula (M57), the Dumbbell (M27) and other bright planetaries such as NGC 6772 in Andromeda or NGC 6543 in Draco. These show several images in different colours, instead of a continuous spectrum as in the stars. They make good subjects for experiments with photography; if colour film is used it will be easy to tell which is the red end of the spectrum. The images due to H-alpha, [OIII], H-beta and H-gamma can be distinguished. Careful inspection will show that the relative intensities of these vary in the different nebulae, reflecting the different physical conditions in each.

5. For work in which you attempt to make intensity measurements, colour film should not be used as its sensitivity varies with wavelength in a complicated way. Use black-and-white film, or a CCD camera.

Appendix 1
Astronomical Societies

For details of local astronomical societies see *International Directory of Astronomical Associations and Societies*, published annually by the Centre de Données de Strasbourg, Université de Strasbourg, or contact your national astronomical society

Agrupación Astronáutica Española,
Rosellón 134,
E-08036 Barcelona,
España

American Association for the Advancement of Science,
1333 II Street NW,
Washington,
DC 2005,
USA

American Association of Variable Star Observers,
25 Birch St,
Cambridge,
MA 02138,
USA

American Astronomical Society,
2000 Florida Avenue NW,
Suite 3000,
Washington,
DC 20009,
USA

Association des Groupes d'Astronomes Amateurs,
4545 Avenue Pierre-de-Coubertin,
Casier Postal 1000, Succ M,
Montreal,
QC H1V 3R2,
Canada

Association Française d'Astronomie,
Observatoire de Montsouris,
17 Rue Emile-Deutsch-de-la-Meurthe,
F-75014 Paris,
France

Association Nationale Science Techniques Jeunesse,
Section Astronomique,
Palais de la Découverte,
Avenue Franklin Roosevelt,
F-75008 Paris,
France

Association of Lunar and Planetary Observers,
PO Box 16131,
San Francisco,
CA 94116,
USA

Astronomical-Geodetical Society of Russia,
24 Sadovaja-Kudrinskaya Ul,
SU-103101 Moskwa,
Russia

Astronomical League,
6235 Omie Circle,
Pensacola,
FL 32504,
USA

Astronomical Society of Australia,
School of Physics,
University of Sydney,
Sydney,
NSW 2006,
Australia

Astronomical Society of Southern Africa.
Southern African Astronomical Observatory,
PO Box 9,
Observatory 7935,
South Africa

Astronomical Society of the Pacific,
1290 24th Avenue,
San Francisco,
CA 94122,
USA

Astronomisk Selskab,
Observatoriet,
Øster Volgade 3,
DK-1350 København K,
Danmark

British Astronomical Association,
Burlington House,
Piccadilly,
London, W1V 9AG,
United Kingdom

British Interplanetary Society,
27/29 South Lambeth Rd,
London, SW8 1SZ,
United Kingdom

Canadian Astronomical Society,
Dominion Astrophysical Observatory,
5071 W. Saanich Rd,
Victoria,
BC V8X 4M6,
Canada

Committee on Space Research (COSPAR),
51 Bd de Montmorency,
F-75016 Paris,
France

Earthwatch,
680 Mount Auburn St,
Box 403,
Watertown,
MA 02272,
USA

Federation of Astronomical Societies,
1 Tal-y-Bont Rd,
Ely,
Cardiff, CF5 5EU,
Wales

International Astronomical Union,
61 Avenue de l'Observatoire,
F-75014 Paris,
France

Junior Astronomical Society,
10 Swanwick Walk,
Tadley,
Basingstoke,
Hampshire, RG26 6JZ,
United Kingdom

National Space Society,
West Wing Suite 203,
600 Maryland Avenue SW,
Washington,
DC 20024,
USA

Nederlandse Astronomenclub,
Netherlands Foundation for Radio Astronomy,
Postbus2,
NL-7990 AA Dwingeloo,
Nederland

Nederlandse Vereniging voor Weer-en Sterrenkunde,
Nachtegaalstrat 82 bis,
NL-3581 AN Utrecht,
Nederland

Nippon Temmon Gakkai,
Tokyo Tenmondai,
2-21-1 Mitaka,
Tokyo 181,
Japan

Royal Astronomical Society,
Burlington House,
Piccadilly,
London, W1V ONL,
United Kingdom

Royal Astronomical Society of Canada,
136 Dupont St,
Toronto, ONT M5R 1V2,
Canada

Royal Astronomical Society of New Zealand,
PO Box 3181,
Wellington,
New Zealand

Schweizerische Astronomische Gesellschaft,
Hirtenhoffstrasse 9,
CH-6005 Luzern,
Switzerland

Società Astronomica Italiana,
Osservatorio Astrofisico di Arcetri,
Largo E. Fermi 5,
I-50125 Firenze,
Italia

Société Astronomique de France,
3 Rue Beethoven,
F-75016 Paris,
France

Société d'Astronomie Populaire,
1 Avenue Camille Flammarion,
F-31500 Toulouse,
France

Société Royale Belge d'Astronomie,
    de Météorologie, et de Physique du Globe,
Observatoire Royale de Belgique,
Avenue Circulaire 3,
B-1180 Bruxelles,
Belgique

Stichting De Koepel,
Nachtegaalstrat 82 bis,
NL-3581 AN Utrecht,
Nederland

Svenska Astronomiska Sallskapet,
Stockholms Observatorium,
S133 00 Saltsjöbaden,
Sweden

Vereinigung der Sternfreunde e.V.,
Volkssternwarte,
Anzingerstrasse 1,
D-8000 Munchen,
Deutschland

Zentral Kommission Astronomie und Raumfahrt,
Postfach 34,
DDR-1030 Berlin,
Deutschland

Appendix 2

Bibliography

A2.1 Journals

Only the major and relatively widely available journals are listed. There are numerous other specialised research-level journals available in academic libraries.

Popular

Astronomy
Astronomy Now
Ciel et Espace
Journal of the British Astronomical Association
New Scientist
Publications of the Astronomical Society of the Pacific
Scientific American
Sky and Telescope

Research

Astronomical Journal
Astronomy and Astrophysics
Astrophysical Journal
Monthly Notices of the Royal Astronomical Association
Nature
Science

A2.2 Ephemerises

Astronomical Almanac, HMSO/US Government Printing Office (published annually).

Handbook of the British Astronomical Association, British Astronomical Association (published annually).
Rocznik Astronomiczny Obserwatorium Krakowskiego, International Supplement, Cracow (Ephemerides of Double Stars).
Yearbook of Astronomy, Macmillan (published annually).

A2.3 Star and Other Catalogues, Atlases and Reference Books

Astrophysical Quantities, CW Allen. Athlone Press, 1973.
Atlas of Representative Stellar Spectra, Y Yamashita, K Nariai and Y Norimoto. University of Tokyo Press, 1977.
Cambridge Deep-Sky Album, J Newton and P Teece. Cambridge University Press, 1983.
Messier Album: An Observer's Handbook, JH Mallas and E Kreimer. Sky Publishing Corporation, 1978.
Messier's Nebulae and Star Clusters, KG Jones. Cambridge University Press, 1991.
Norton's 2000.0, I Ridpath (Ed.). Longman, 1989.
Observing Handbook and Catalogue of Deep Sky Objects, C Luginbuhl and B Skiff. Cambridge University Press, 1990.
Planetary Nebulae: A Practical Guide and Handbook for Amateur Astronomers, SJ Hynes. Willmann-Bell, 1991.
Sky Atlas 2000.0, W Tirion. Sky Publishing Corporation, 1981.
Sky Catalogue 2000, Volumes 1 and 2, A Hirshfeld and RW Sinnott. Cambridge University Press, 1985.
Times Atlas of the Moon, HAG Lewis (Ed.). Times Newspapers, 1969.
URANOMETRIA 2000.0, W Tirion, B Rappaport and G Lovi. Willmann-Bell, 1988.

A2.4 Introductory Astronomy Books

Astronomy: A Self-Teaching Guide, DL Moche. John Wiley, 1993.

Astronomy: Principles and Practice, AE Roy and D Clark. Adam Hilger, 1988.

Astronomy: The Evolving Universe, M Zeilik. John Wiley, 1994.

Astronomy through Space and Time, S. Engelbrektson. WC Brown, 1994.

Introductory Astronomy and Astrophysics, M Zeilik, SA Gregory and EvP Smith. Saunders, 1992.

Universe, WJ Kaufmann III. WH Freeman, 1994.

A2.5 Practical Astronomy Books

Amateur Astronomer's Handbook, JB Sidgwick. Faber & Faber, 1971.

Analysis of Starlight: 150 Years of Astronomical Spectroscopy, JB Hearnshaw. Cambridge University Press, 1987.

Astronomical Photometry: A Guide, C Sterken and J Manfroid. Kluwer, 1992.

Astronomical Spectroscopy, C Kitchin. Adam Hilger, 1995.

Astronomical Telescope, BV Barlow. Wykeham Publications, 1975.

Astronomy on the Personal Computer, O Montenbruck and T Pfleger. Springer-Verlag, 1991.

Astrophysical Techniques, CR Kitchin. Adam Hilger, 1991.

Beginner's Guide to Astronomical Telescope Making, J Muirden. Pelham Press, 1975.

Building and Using an Astronomical Observatory, P Doherty. Stevens, 1986.

CCD Astronomy, C Buil. Willmann-Bell, 1991.

Challenges of Astronomy, W Schlosser, T Schmidt-Kaler and EF Malone. Springer-Verlag, 1991.

Compendium of Practical Astronomy, GD Roth. Springer-Verlag, 1993.

Computer Processing of Remotely-Sensed Images: An Introduction, PW Mather. John Wiley, 1987.

Data Analysis in Astronomy, V Di Gesu, L Scarsi and R Buccheri. Plenum Press, 1992.

Exercises in Practical Astronomy using Photographs, MT Bruck. Adam Hilger, 1990.

Getting the Measure of Stars, WA Cooper and EN Walker. Adam Hilger, 1989.

Handbook for Telescope Making, NE Howard. Faber & Faber, 1962.

Introduction to Astronomical Image Processing, R Berry. Willmann-Bell, 1991.

Introduction to Experimental Astronomy, RB Culver. WH Freeman, 1984.

Manual of Advanced Celestial Photography, BD Wallis and RW Provin. Cambridge University Press, 1988.

Mathematical Astronomy with a Pocket Calculator, A. Jones. David and Charles, 1978.

The Modern Amateur Astronomer, P Moore (Ed.). Springer-Verlag, 1995.

Observational Astronomy, DS Birney. Cambridge University Press, 1991.

Observing the Sun, PO Taylor. Cambridge University Press, 1991.

Practical Astronomer, CA Ronan. Pan, 1981.

Practical Astronomy with your Calculator, P Duffett-Smith. Cambridge University Press, 1981.

Practical Astronomy: A User Friendly Handbook for Skywatchers, HR Mills. Albion, 1993.

Seeing the Sky: 100 Projects, Activities and Explorations in Astronomy, F. Schaaf. John Wiley, 1990.

Small Astronomical Observatories, P Moore (Ed.). Springer-Verlag, 1996.

Solar System: A Practical Guide, D. Reidy and K Wallace. Allen and Unwin, 1991.

Star Gazing through Binoculars: A Complete Guide to Binocular Astronomy, S. Mensing. TAB, 1986.

Star Hopping: Your Visa to the Universe, R.A. Garfinkle. Cambridge University Press, 1993.

Sundials, RN Mayall and ML Mayall. Charles T Branford, 1958.

Telescopes and Techniques, C Kitchin. Springer-Verlag, 1995.

Workbook for Astronomy, J Waxman. Cambridge University Press, 1984.

Appendix 3
Messier and Caldwell Catalogues

Charles Messier (1730–1817) was an observational astronomer working from Paris in the eighteenth century. He discovered between 15 and 21 comets and observed many more. During his observations he encountered nebulous objects which were not comets. Some of these objects were his own discoveries, while others had been known before. In 1774 he published a list of 45 of these nebulous objects. His purpose in publishing the list was so that other comet-hunters should not confuse the nebulae with comets. Over the following decades he published supplements which increased the number of objects in his catalogue to 103, though objects M101 and M102 were in fact the same. Later other astronomers added a replacement for M102 and objects 104 to 110. It is now thought probable that Messier had observed these later additions with the exception of the last. Thus the modern version of his catalogue has 109 objects in it. Several of the objects in Messier's original lists are now difficult to identify, and "best guesses" have had to be made regarding the objects intended. Ironically in one case (M91) it is possible that Messier's original observation was of an unrecognised comet!

Messier observed mostly with 3–3.5-inch (75–90 mm) refractors. He had access to a 7.5-inch (190 mm) Gregorian reflector but since this used mirrors made from speculum metal, its equivalent aperture would have been only 3 inches or so. With modern telescopes it should be possible to observe all the Messier objects with a 2.5-inch (60 mm) or larger instrument. This accessibility of the objects in Messier's list, compared with *Herschel's New General Catalogue* (NGC) which was being compiled at the same time as Messier's observations but using much larger telescopes, probably explains its modern popularity. It is a challenging but achievable task for most amateur astronomers to observe all the Messier objects. At "star parties" and within astronomy clubs, going for the maximum number of Messier objects observed is a popular competition. Indeed, at some times of the year it is just about possible to observe most of them in a single night.

Messier observed from Paris and therefore the most southerly object in his list is M7 in Scorpius with a declination of –35°. He also missed several objects from his list, such as h and χ Per and the Hyades, which most observers would feel should have been included. The well-known astronomer Dr Patrick Moore has therefore recently introduced the *Caldwell Catalogue* (his full name is Patrick *Caldwell*-Moore). This has 109 objects like Messier's list, but covers the whole sky. The Caldwell objects are listed in decreasing order of declination, so that from a given latitude all objects from C1 to Cn (or from C109 to Cn for southern observers), where n is the number of the most southerly (or northerly) object rising at that site, should be visible. There is no overlap between the *Messier* and *Caldwell Catalogues*, and the two taken together (see below) will give sufficient fascinating and spectacular objects to keep most astronomers occupied for several years' worth of observing.

Table A3.1.   The Messier Objects

Object	Nebula name	NGC	Type	RA$_{2000}$ H	m	Dec$_{2000}$ degrees	Constellation	Visual[a] mag.	Size[b] (')
Messier 1	Crab	1952	Supernova Remnant	05	35	22.0	Tau	8.4	6
Messier 2		7089	Globular Cluster	21	34	−0.8	Aqr	6.5	13
Messier 3		5272	Globular Cluster	13	42	28.4	CVn	6.4	16
Messier 4		6121	Globular Cluster	16	24	−26.5	Sco	5.9	26
Messier 5		5904	Globular Cluster	15	19	2.1	Ser Cap	5.8	17
Messier 6	Bufferfly	6405	Open Cluster	17	40	−32.2	Sco	4.2	15
Messier 7		6475	Open Cluster	17	54	−34.8	Sco	3.3	80
Messier 8	Lagoon	6523	Emission Nebula	18	04	−24.4	Sgr	5.8	90
Messier 9		6333	Globular Cluster	17	19	−18.5	Oph	7.9	9
Messier 10		6254	Globular Cluster	16	57	−4.1	Oph	6.6	15
Messier 11	Wild Duck	6705	Open Cluster	18	51	−6.3	Sct	5.8	14
Messier 12		6218	Globular Cluster	16	47	−2.0	Oph	6.6	15
Messier 13		6205	Globular Cluster	16	42	36.5	Her	5.9	17
Messier 14		6402	Globular Cluster	17	38	−3.3	Oph	7.6	12
Messier 15		7078	Globular Cluster	21	30	12.2	Peg	6.4	12
Messier 16	Eagle	6611	Open Cluster	18	19	−13.8	Ser	6.0	7
Messier 17	Omega	6618	Emission Nebula	18	21	−16.2	Sgr	6.0	46
Messier 18		6613	Open Cluster	18	20	−17.1	Sgr	6.9	9
Messier 19		6273	Globular Cluster	17	03	−26.3	Oph	7.2	14
Messier 20	Trifid	6514	Emission Nebula	18	03	−23.0	Sgr	6.3	29
Messier 21		6531	Open Cluster	18	05	−22.5	Sgr	5.9	13
Messier 22		6656	Globular Cluster	18	36	−23.9	Sgr	5.1	24
Messier 23		6494	Open Cluster	17	57	−19.0	Sgr	5.5	27
Messier 24		6603	Open Cluster	18	18	−18.4	Sgr	11.1	5
Messier 25		IC 4725	Open Cluster	18	32	−19.3	Sgr	4.6	32
Messier 26		6694	Open Cluster	18	45	−9.4	Sct	8.0	15
Messier 27	Dumbbell	6853	Planetary Nebula	20	00	22.7	Vul	7.6	15
Messier 28		6626	Globular Cluster	18	25	−24.9	Sgr	6.9	11
Messier 29		6913	Open Cluster	20	24	38.5	Cyg	6.6	7
Messier 30		7099	Globular Cluster	21	40	−23.2	Cap	7.5	11
Messier 31	Andromeda	224	Galaxy	00	43	41.3	And	3.5	180
Messier 32		221	Galaxy	00	43	40.9	And	8.2	8
Messier 33	Triangulum	598	Galaxy	01	34	30.7	Tri	5.7	62
Messier 34		1039	Open Cluster	02	42	42.8	Per	5.2	35
Messier 35		2168	Open Cluster	06	09	24.3	Gem	5.1	28
Messier 36		1960	Open Cluster	05	36	34.1	Aur	6.0	12
Messier 37		2099	Open Cluster	05	52	32.6	Aur	5.6	24
Messier 38		1912	Open Cluster	05	29	35.8	Aur	6.4	21
Messier 39		7092	Open Cluster	21	32	48.4	Cyg	4.6	32
Messier 40			2 stars	12	22	58.1	UMa	0.8	9

[a] This is the integrated magnitude over the whole area of the object. An angularly large object with a bright magnitude may therefore be less easy to see than a smaller object with a fainter magnitude. The magnitudes of the emission nebulae in particular may be misleading because they frequently contain brighter and darker regions.

[b] This is the largest dimension of the object. Some objects may be filamentary or have a brighter core or outer region, making them easier to see than might be expected.

Table A3.1 (continued)

Object	Nebula name	NGC	Type	RA2000 H	RA2000 m	Dec2000 degrees	Constellation	Visual[a] mag.	Size[b] (')
Messier 41		2287	Open Cluster	06	47	−20.7	CMa	4.5	38
Messier 42	Orion	1976	Emission Nebula	05	35	−5.5	Ori	4.0	66
Messier 43	Orion	1982	Emission Nebula	05	36	−5.3	Ori	9.0	20
Messier 44	Praesepe	2632	Open Cluster	08	40	20.0	Cnc	3.1	95
Messier 45	Pleiades		Open Cluster	03	47	24.1	Tau	1.2	110
Messier 46		2437	Open Cluster	07	42	−14.8	Pup	6.1	27
Messier 47		2422	Open Cluster	07	37	−14.5	Pup	4.4	30
Messier 48		2548	Open Cluster	08	14	−5.8	Hya	5.8	54
Messier 49		4472	Galaxy	12	30	8.0	Vir	8.4	9
Messier 50		2323	Open Cluster	07	03	−8.3	Mon	5.9	16
Messier 51	Whirlpool	5194	Galaxy	13	30	47.2	CVn	8.4	11
Messier 52		7654	Open Cluster	23	24	61.6	Cas	6.9	13
Messier 53		5024	Globular Cluster	13	13	18.2	Com	7.7	13
Messier 54		6715	Globular Cluster	18	55	−30.5	Sgr	7.7	9
Messier 55		6809	Globular Cluster	19	40	−31.0	Sgr	7.0	19
Messier 56		6779	Globular Cluster	19	17	30.2	Lyr	8.3	7
Messier 57	Ring	6720	Planetary Nebula	18	54	33.0	Lyr	9.7	3
Messier 58		4579	Galaxy	12	38	11.8	Vir	9.8	5
Messier 59		4621	Galaxy	12	42	11.7	Vir	9.8	5
Messier 60		4649	Galaxy	12	44	11.6	Vir	8.8	7
Messier 61		4303	Galaxy	12	22	4.5	Vir	9.7	6
Messier 62		6266	Globular Cluster	17	01	−30.1	Oph	6.6	14
Messier 63	Sunflower	5055	Galaxy	13	16	42.0	CVn	8.6	12
Messier 64	Black-eye	4826	Galaxy	12	57	21.7	Com	8.5	9
Messier 65		3623	Galaxy	11	19	13.1	Leo	9.3	10
Messier 66		3627	Galaxy	11	20	13.0	Leo	9.0	9
Messier 67		2682	Open Cluster	08	50	11.8	Cnc	6.9	30
Messier 68		4590	Globular Cluster	12	40	−26.8	Hya	8.2	12
Messier 69		6637	Globular Cluster	18	31	−32.4	Sgr	7.7	7
Messier 70		6681	Globular Cluster	18	43	−32.3	Sgr	8.8	8
Messier 71		6838	Globular Cluster	19	54	18.8	Sge	8.3	7
Messier 72		6981	Globular Cluster	20	54	−12.5	Aqr	9.4	6
Messier 73		6994	Open Cluster	20	59	−12.6	Aqr	8.9	3
Messier 74		628	Galaxy	01	37	15.8	Psc	9.2	10
Messier 75		6864	Globular Cluster	20	06	−21.9	Sgr	8.6	6
Messier 76	Little Dumbbell	650	Planetary Nebula	01	42	51.6	Per	12.2	5
Messier 77		1068	Galaxy	02	43	0.0	Cet	8.8	7
Messier 78		2068	Reflection Nebula	05	47	0.1	Ori	8.0	8
Messier 79		1904	Globular Cluster	05	25	−24.6	Lep	8.0	9
Messier 80		6093	Globular Cluster	16	17	−23.0	Sco	7.2	9
Messier 81	Bode's	3031	Galaxy	09	56	69.1	UMa	6.9	26
Messier 82		3034	Galaxy	09	56	69.7	UMa	8.4	11
Messier 83		5236	Galaxy	13	37	−29.9	Hya	8.2	11
Messier 84		4374	Galaxy	12	25	12.9	Vir	9.3	5
Messier 85		4382	Galaxy	12	25	18.2	Com	9.2	7

(continued overleaf)

Table A3.1 (*continued*)

Object	Nebula name	NGC	Type	RA$_{2000}$ H	m	Dec$_{2000}$ degrees	Constellation	Visual[a] mag.	Size[b] (')
Messier 86		4406	Galaxy	12	26	13.0	Vir	9.2	7
Messier 87	Virgo A	4486	Galaxy	12	31	12.4	Vir	8.6	7
Messier 88		4501	Galaxy	12	32	14.4	Com	9.5	7
Messier 89		4552	Galaxy	12	36	12.6	Vir	9.8	4
Messier 90		4569	Galaxy	12	37	13.2	Vir	9.5	10
Messier 91		4548	Galaxy	12	35	14.5	Com	10.2	5
Messier 92		6341	Globular Cluster	17	17	43.1	Her	6.5	11
Messier 93		2447	Open Cluster	07	45	−23.9	Pup	6.2	22
Messier 94		4736	Galaxy	12	51	41.1	CVn	8.2	11
Messier 95		3351	Galaxy	10	44	11.7	Leo	9.7	7
Messier 96		3368	Galaxy	10	47	11.8	Leo	9.2	7
Messier 97	Owl	3587	Planetary Nebula	11	15	55.0	UMa	12	3
Messier 98		4192	Galaxy	12	14	14.9	Com	10.1	10
Messier 99		4254	Galaxy	12	19	14.4	Com	9.8	5
Messier 100		4321	Galaxy	12	23	15.8	Com	9.4	7
Messier 101	Pinwheel	5457	Galaxy	14	03	54.4	UMa	7.7	27
Messier 102		5866	Galaxy	15	07	55.8	Dra	10.0	5
Messier 103		581	Open Cluster	01	33	60.7	Cas	7.4	6
Messier 104	Sombrero	4594	Galaxy	12	40	−11.6	Vir	8.3	9
Messier 105		3379	Galaxy	10	48	12.6	Leo	9.3	5
Messier 106		4258	Galaxy	12	19	47.3	CVn	8.3	18
Messier 107		6171	Globular Cluster	16	33	−13.1	Oph	8.1	10
Messier 108		3556	Galaxy	11	12	55.7	UMa	10.1	8
Messier 109		3992	Galaxy	11	58	53.4	UMa	9.8	8

Data for this table obtained from *Sky Catalogue 2000.0*, Vol. 2 (Ed. A. Hirshfeld and R.W. Sinnott, Cambridge University Press, 1985); *Astrophysical Quantities* (C.W. Allen, Athlone Press, 1973); *NGC 2000.0* (R.W. Sinnott, Cambridge University Press, 1988); *Visual Astronomy of the Deep Sky* (R.N. Clark, Cambridge University Press, 1990); *Hartung's Astronomical Objects for Southern Telescopes* (D. Malin and, D.J. Frew, Cambridge University Press, 1995); and *Astrophysical Journal Supplement*, **4**, 257, 1959, S. Sharpless.

Table A3.2. The Caldwell Objects

Object	Nebula name	NGC	Type	RA2000 H	m	Dec2000 degrees	Constellation	Visual[a] mag.	Size[b] (')
Caldwell 1		188	Open Cluster	00	44	85.3	Cep	8.1	14
Caldwell 2		40	Planetary Nebula	00	13	72.5	Cep	10.7	0.6
Caldwell 3		4236	Galaxy	12	17	69.5	Dra	9.7	19
Caldwell 4		7023	Reflection Nebula	21	02	68.2	Cep	7.0	18
Caldwell 5		IC 342	Galaxy	03	47	68.1	Cam	9.1	18
Caldwell 6	Cat's Eye	6543	Planetary Nebula	17	59	66.6	Dra	8.8	6
Caldwell 7		2403	Galaxy	07	37	65.6	Cam	8.4	18
Caldwell 8		559	Open Cluster	01	30	63.3	Cas	9.5	5
Caldwell 9	Cave	Sh2-155	Emission Nebula	22	57	62.6	Cep	≈ 9	50
Caldwell 10		663	Open Cluster	01	46	61.3	Cas	7.1	16
Caldwell 11	Bubble	7635	Emission Nebula	23	21	61.2	Cas	8.5	15
Caldwell 12		6946	Galaxy	20	35	60.2	Cep	8.9	11
Caldwell 13		457	Open Cluster	01	19	58.3	Cas	6.4	13
Caldwell 14	h and chi Per	869/884	Open Cluster	02	20	57.1	Per	4.3/4.4	30/30
Caldwell 15	Blinking	6826	Planetary Nebula	19	45	50.5	Cyg	9.8	2
Caldwell 16		7243	Open Cluster	22	15	49.9	Lac	6.4	21
Caldwell 17		147	Galaxy	00	33	48.5	Cas	9.3	13
Caldwell 18		185	Galaxy	00	39	48.3	Cas	9.2	12
Caldwell 19	Cocoon	IC 5146	Emission Nebula	21	54	47.3	Cyg	7.2	12
Caldwell 20	North America	7000	Emission Nebula	20	59	44.3	Cyg	5.0	120
Caldwell 21		4449	Galaxy	12	28	44.1	CVn	9.4	5
Caldwell 22	Blue Snowball	7662	Planetary Nebula	23	26	42.6	And	9.2	2
Caldwell 23		891	Galaxy	02	23	42.4	And	10.0	14
Caldwell 24		1275	Galaxy	03	20	41.5	Per	11.6	3
Caldwell 25		2419	Globular Cluster	07	38	38.9	Lyn	10.4	4
Caldwell 26		4244	Galaxy	12	18	37.8	CVn	10.2	16
Caldwell 27	Crescent	6888	Emission Nebula	20	12	38.4	Cyg	≈ 11	20
Caldwell 28		752	Open Cluster	01	58	37.7	And	5.7	50
Caldwell 29		5005	Galaxy	13	11	37.1	CVn	9.8	5
Caldwell 30		7331	Galaxy	22	37	34.4	Peg	9.5	11
Caldwell 31	Flaming Star	IC 405	Emission Nebula	05	16	34.3	Aur	≈ 7	30
Caldwell 32		4631	Galaxy	12	42	32.5	CVn	9.3	15
Caldwell 33	Veil (E)	6992/5	Supernova Remnant	20	57	31.5	Cyg	8.0	60
Caldwell 34	Veil (W)	6960	Supernova Remnant	20	46	30.7	Cyg	8.0	70
Caldwell 35		4889	Galaxy	13	00	28.0	Com	11.4	3
Caldwell 36		4559	Galaxy	12	36	28.0	Com	9.9	11
Caldwell 37		6885	Open Cluster	20	12	26.5	Vul	5.7	7
Caldwell 38		4565	Galaxy	12	36	26.0	Com	9.6	16
Caldwell 39	Eskimo	2392	Planetary Nebula	07	29	20.9	Gem	9.9	0.7
Caldwell 40		3626	Galaxy	11	20	18.4	Leo	10.9	3

[a] This is the integrated magnitude over the whole area of the object. An angularly large object with a bright magnitude may therefore be less easy to see than a smaller object with a fainter magnitude. The magnitudes of the emission nebulae in particular may be misleading because they frequently contain brighter and darker regions. The symbol "≈" indicates a magnitude estimated from visual descriptions.
[b] This is the largest dimension of the object. Some objects may be filamentary or have a brighter core or outer region, making them easier to see than might be expected.

(continued overleaf)

Table A3.2 (*continued*)

Object	Nebula name	NGC	Type	RA₂₀₀₀ H	m	Dec₂₀₀₀ degrees	Constellation	Visual[a] mag.	Size[b] (')
Caldwell 41	Hyades		Open Cluster	04	27	16.0	Tau	0.5	330
Caldwell 42		7006	Globular Cluster	21	02	16.2	Del	10.6	3
Caldwell 43		7814	Galaxy	00	03	16.2	Peg	10.5	6
Caldwell 44		7479	Galaxy	23	05	12.3	Peg	11.0	4
Caldwell 45		5248	Galaxy	13	38	8.9	Boö	10.2	7
Caldwell 46	Hubble's variable	2261	Reflection Nebula	06	39	8.7	Mon	10.0	2
Caldwell 47		6934	Globular Cluster	20	34	7.4	Del	8.9	6
Caldwell 48		2775	Galaxy	09	10	7.0	Cnc	10.3	5
Caldwell 49	Rosette	2237-9	Emission Nebula	06	32	5.1	Mon	≈ 4	80
Caldwell 50		2244	Open Cluster	06	32	4.9	Mon	4.8	24
Caldwell 51		IC 1613	Galaxy	01	05	2.1	Cet	9.3	12
Caldwell 52		4697	Galaxy	12	49	−5.8	Vir	9.3	6
Caldwell 53	Spindle	3115	Galaxy	10	05	−7.7	Sex	9.2	8
Caldwell 54		2506	Open Cluster	08	00	−10.8	Mon	7.6	7
Caldwell 55	Saturn	7009	Planetary Nebula	21	04	−11.4	Aqr	8.3	2
Caldwell 56		246	Planetary Nebula	00	47	−11.9	Cet	8.0	4
Caldwell 57	Barnard's	6822	Galaxy	19	45	−14.8	Sgr	9.4	10
Caldwell 58		2360	Open Cluster	07	18	−15.6	CMa	7.2	13
Caldwell 59	Ghost of Jupiter	3242	Planetary Nebula	10	25	−18.6	Hya	8.6	21
Caldwell 60	Antennae	4038	Galaxy	12	02	−18.9	Crv	10.7	3
Caldwell 61	Antennae	4039	Galaxy	12	02	−18.9	Crv	10.7	3
Caldwell 62		247	Galaxy	00	47	−20.8	Cet	8.9	20
Caldwell 63	Helix	7293	Planetary Nebula	22	30	−20.8	Aqr	7.4	13
Caldwell 64		2362	Open Cluster	07	19	−25.0	CMa	4.1	8
Caldwell 65	Silver Coin	253	Galaxy	00	48	−25.3	Scl	7.1	25
Caldwell 66		5694	Globular Cluster	14	40	−26.5	Hya	10.2	4
Caldwell 67		1097	Galaxy	02	46	−30.3	For	9.3	9
Caldwell 68	R CrA	6729	Reflection Nebula	19	02	−37.0	CrA	≈ 11	1
Caldwell 69	Bug	6302	Planetary Nebula	17	14	−37.1	Sco	12.8	1
Caldwell 70		300	Galaxy	00	55	−37.7	Scl	8.7	20
Caldwell 71		2477	Open Cluster	07	52	−38.6	Pup	5.8	27
Caldwell 72		55	Galaxy	00	15	−39.2	Scl	7.9	32
Caldwell 73		1851	Globular Cluster	05	14	−40.1	Col	7.3	11
Caldwell 74	Eight-Burst	3132	Planetary Nebula	10	08	−40.4	Vel	8.2	0.8
Caldwell 75		6124	Open Cluster	16	26	−40.7	Sco	5.8	29
Caldwell 76		6231	Open Cluster	16	54	−41.8	Sco	2.6	15
Caldwell 77	Cen A	5128	Galaxy	13	26	−43.0	Cen	7.0	18
Caldwell 78		6541	Globular Cluster	18	08	−43.7	CrA	6.6	13
Caldwell 79		3201	Globular Cluster	10	18	−46.4	Vel	6.8	18
Caldwell 80	Omega Centauri	5139	Globular Cluster	13	27	−47.5	Cen	3.7	36
Caldwell 81		6352	Globular Cluster	17	26	−48.4	Ara	8.2	7
Caldwell 82		6193	Open Cluster	16	41	−48.8	Ara	5.2	15
Caldwell 83		4945	Galaxy	13	05	−49.5	Cen	8.6	20
Caldwell 84		5286	Globular Cluster	13	46	−51.4	Cen	7.6	9
Caldwell 85		IC 2391	Open Cluster	08	40	−53.1	Vel	2.5	50

Table A3.2 (*continued*)

Object	Nebula name	NGC	Type	RA2000 H	RA2000 m	Dec2000 degrees	Constellation	Visual[a] mag.	Size[b] (')
Caldwell 86		6397	Globular Cluster	17	41	−53.7	Ara	5.7	26
Caldwell 87		1261	Globular Cluster	03	12	−55.2	Hor	8.4	7
Caldwell 88		5823	Open Cluster	15	06	−55.6	Cir	7.9	10
Caldwell 89	S Norma	6087	Open Cluster	16	19	−57.9	Nor	5.4	12
Caldwell 90		2867	Planetary Nebula	09	21	−58.3	Car	9.7	0.2
Caldwell 91		3532	Open Cluster	11	06	−58.7	Car	3.0	55
Caldwell 92	Eta Carina	3372	Emission Nebula	10	44	−59.9	Car	2.5	120
Caldwell 93		6752	Globular Cluster	19	11	−60.0	Pav	5.4	20
Caldwell 94	Jewel Box	4755	Open Cluster	12	54	−60.3	Cru	4.2	10
Caldwell 95		6025	Open Cluster	16	04	−60.5	TrA	5.1	12
Caldwell 96		2516	Open Cluster	07	58	−60.9	Car	3.8	30
Caldwell 97		3766	Open Cluster	11	36	−61.6	Cen	5.3	12
Caldwell 98		4609	Open Cluster	12	42	−63.0	Cru	6.9	5
Caldwell 99	Coalsack		Absorption Nebula	12	53	−63.0	Cru	−	350
Caldwell 100		IC 2944	Open Cluster	11	37	−63.0	Cen	4.5	15
Caldwell 101		6744	Galaxy	19	10	−63.9	Pav	8.4	16
Caldwell 102	Southern Pleiades	IC 2602	Open Cluster	10	43	−64.4	Car	1.9	50
Caldwell 103	Tarantula	2070	Emission Nebula	05	39	−69.1	Dor	8.2	40
Caldwell 104		362	Globular Cluster	01	03	−70.9	Tuc	6.6	13
Caldwell 105		4833	Globular Cluster	13	00	−70.9	Mus	7.4	14
Caldwell 106	47 Tucanae	104	Globular Cluster	00	24	−72.1	Tuc	4.0	31
Caldwell 107		6101	Globular Cluster	16	26	−72.2	Aps	9.3	11
Caldwell 108		4372	Globular Cluster	12	26	−72.7	Mus	7.8	19
Caldwell 109		3195	Planetary Nebula	10	10	−80.9	Cha	11.6	0.6

Data for this table obtained from *Sky Catalogue 2000.0*, Vol. 2 (Ed. A. Hirshfeld and R.W. Sinnott, Cambridge University Press, 1985); *The Caldwell Card* (Sky Publishing Corp., 1996); *NGC 2000.0* (R.W. Sinnott, Cambridge University Press, 1988); *Visual Astronomy of the Deep Sky* (R.N. Clark, Cambridge University Press, 1990); *Hartung's Astronomical Objects for Southern Telescopes* (D. Malin and D.J. Frew, Cambridge University Press, 1995); and *Astrophysical Journal Supplement*, **4**, 257, 1959, S. Sharpless.

Appendix 4

A Selection of Choice Astronomical Objects for Viewing

Table A4.1. Galactic Clusters

Cluster	NGC	RA$_{2000}$ (h	m)	Dec$_{2000}$ (degree)	m$_v$[a]	Angular size (')	Number of stars[b]
M103	581	01	33	+60.7	7.4	6	25
h Per (C14)	869	02	19	+57.2	4.3	30	200
χ Per (C14)	884	02	22	+57.1	4.4	30	150
M34	1039	02	42	+42.8	5.2	35	60
Pleiades (M45)		03	47	+24.1	1.2	2°	100
Hyades (C41)		04	27	+16	0.5	5.5°	100
M38	1912	05	29	+35.8	6.4	21	100
M36	1960	05	36	+34.1	6.0	12	60
Cr70		05	36	–01	0.4	2.5°	100
M37	2099	05	52	+32.6	5.6	24	150
M35	2168	06	09	+24.3	5.1	28	200
Rosette (C50)	2244	06	32	+04.9	4.8	24	100
M41	2287	06	47	–20.7	4.5	38	80
M50	2323	07	03	–8.3	5.9	16	80
τ CMa (C64)	2362	07	19	–25.0	4.1	8	60
M46	2437	07	42	–14.8	6.1	27	100
M93	2447	07	45	–23.9	6.2	22	80
C71	2477	07	52	–38.6	5.8	27	160
C96	2516	07	58	–60.9	3.8	30	80
	2547	08	11	–49.3	4.7	20	80
M48	2548	08	14	–05.8	5.8	54	80
Praesepe (M44)	2632	08	40	+20.0	3.1	1.5°	50
o Vel		08	40	–53.1	2.5	50	30
θ Car		10	43	–64.4	1.9	50	60
C91	3532	11	06	–58.7	3.0	55	150

[a] This is the magnitude of all the stars in the cluster added together, sometimes called the integrated magnitude.
[b] This is the number of stars discernible on the Palomar Sky Survey plates. Fewer stars are likely to be visible in a small telescope. Detailed studies with large telescopes, however, may show many more members of the cluster; up to 3000 for example in the Pleiades.

(continued overleaf)

Table A4.1 (*continued*)

Cluster	NGC	RA$_{2000}$ (h)	(m)	Dec$_{2000}$ (degree)	m$_v$[a]	Angular size (')	Number of stars[b]
C97	3766	11	36	−61.6	5.3	12	100
Ursa Major		12		+60	0.4	28°	100
Coma		12	25	+26	1.8	4.5°	80
Jewel Box (C94)	4755	12	54	−60.3	4.2	10	30
	5316	13	54	−61.9	6.0	14	80
	5662	14	35	−56.6	5.5	12	70
C95	6025	16	04	−60.5	5.1	12	60
C89 (S Norma)	6087	16	14	−57.9	5.4	12	100
C75	6124	16	26	−40.7	5.8	29	100
	6250	16	58	−45.8	5.9	8	60
Butterfly (M6)	6405	17	40	−32.2	4.2	15	80
M7	6475	17	54	−34.8	3.3	1.4°	80
M23	6494	17	57	−19.0	5.5	27	150
M21	6531	18	05	−22.5	5.9	13	70
M16	6611	18	19	−13.8	6.0	7	100
Omega (M17)	6618	18	21	−16.2	6.0	11	40
M25		18	32	−19.3	4.6	32	30
IC4756		18	39	+05.5	5.4	52	80
M26	6694	18	45	−09.4	8.0	15	30
M11 (Wild Duck)	6705	18	51	−06.3	5.8	14	200
M29	6913	20	24	+38.5	6.6	7	50
M39	7092	21	32	+48.4	4.6	32	30
M52	7654	23	24	+61.6	6.9	13	100

[a] This is the magnitude of all the stars in the cluster added together, sometimes called the integrated magnitude.

[b] This is the number of stars discernible on the Palomar Sky Survey plates. Fewer stars are likely to be visible in a small telescope. Detailed studies with large telescopes, however, may show many more members of the cluster; up to 3000 for example in the Pleiades.

Table A4.2. Globular Clusters

Cluster	NGC	RA₂₀₀₀ (h)	(m)	Dec₂₀₀₀ (°)	m_v^a	Angular size (')
47 Tuc (C106)	104	00	24	–72.1	4.0	31
C104	362	01	03	–70.9	6.6	13
	2808	09	12	–64.9	6.3	14
C79	3201	10	18	–46.4	6.8	18
M68	4590	12	40	–26.8	8.2	12
ω Cen (C80)	5139	13	27	–47.5	3.7	36
M3	5272	13	42	+28.4	6.4	16
M5	5904	15	19	+02.1	5.8	17
M80	6093	16	17	–23.0	7.2	9
M4	6121	16	24	–26.5	5.9	26
M13	6205	16	42	+36.5	5.9	17
M12	6218	16	47	–02.0	6.6	15
M10	6254	16	57	–04.1	6.6	15
M62	6266	17	01	–30.1	6.6	14
M19	6273	17	03	–26.3	7.2	14
M92	6341	17	17	+43.1	6.5	11
M9	6333	17	19	–18.5	7.9	9
	6388	17	36	–44.7	6.9	9
M14	6402	17	38	–03.3	7.6	12
C86	6397	17	41	–53.7	5.7	26
C78	6541	18	08	–43.7	6.6	13
M28	6626	18	25	–24.9	6.9	11
M69	6637	18	31	–32.4	7.7	7
M22	6656	18	36	–23.9	5.1	24
M70	6681	18	43	–32.3	8.1	8
M54	6715	18	55	–30.5	7.7	9
C93	6752	19	11	–60.0	5.4	20
M56	6779	19	17	+30.2	8.3	7
M55	6809	19	40	–31.0	7.0	19
M71	6838	19	54	+18.8	8.3	7
M75	6864	20	06	–21.9	8.6	6
M72	6981	20	54	–12.5	9.4	6
M15	7078	21	30	+12.2	6.4	12
M2	7089	21	34	–00.8	6.5	13
M30	7099	21	40	–23.2	7.5	11

^{a} This is the magnitude of all the stars in the cluster added together, sometimes called the integrated magnitude.

Table A4.3. Nebulae

Nebula	NGC	RA$_{2000}$ (h	m)	Dec$_{2000}$ (')	Angular size (')	Type
C2	40	00	13	+72.5	0.6	Planetary
Little Dumbbell, M76	650/1	01	42	+51.6	1 × 4	Planetary
Merope nebulosity	1435	03	46	+24.6	30 × 30	Reflection
California Nebula	1499	04	01	+36.6	145 × 40	Emission
Flaming Star Nebula, C31	1405	05	16	+34.3	30 × 19	Emission
Crab Nebula, M1	1952	05	35	+22.0	6 × 4	Emission
Orion Nebula, M42	1976	05	35	−05.5	66 × 60	Emission
Tarantula Nebula, C103	2070	05	39	−69.1	40 × 25	Emission
Horsehead Nebula		05	41	−02.5	6 × 4	Dark
	2024	05	41	−02.5	30 × 30	Emission
M78	2068	05	47	+00.1	8 × 6	Reflection
Rosette Nebula, C49	2237	06	32	+05.1	80 × 60	Emission
Hubble's Variable Nebula, C46	2261	06	39	+08.7	2 × 1	Reflection
Eskimo Nebula, C39	2392	07	29	+20.9	0.2	Planetary
C90	2867	09	21	−58.3	0.2	Planetary
C74	3132	10	08	−40.4	0.8	Planetary
C109	3195	10	10	−80.9	0.6	Planetary
Ghost of Jupiter, C59	3242	10	25	−18.6	0.3	Planetary
Eta Carinae Nebula, C92	3372	10	44	−59.9	120 × 120	Emission
Owl Nebula, M97	3587	11	15	+55.0	3	Planetary
The Coalsack, C99		12	53	−63	400 × 300	Dark
Rho Ophiuchi	I4604	16	26	−23.4	60 × 25	Reflection
Bug Nebula, C69	6302	17	14	−37.1	0.8	Planetary
Pipe Nebula (stem)		17	21	−27	300 × 60	Dark
Snake Nebula		17	24	−23.6	4	Dark
Pipe Nebula (bowl)		17	33	−26	200 × 140	Dark
Cat's Eye Nebula, C6	6543	17	59	+66.6	0.3	Planetary
Trifid Nebula, M20	6514	18	03	−23.0	29 × 27	Emis. + Refl.
Lagoon Nebula, M8	6523	18	04	−24.4	90 × 40	Emission
Eagle Nebula, M16	I4703	18	19	−13.8	35 × 28	Emission
Omega Nebula, M17	6618	18	21	−16.2	46 × 37	Emission
Ring Nebula, M57	6720	18	54	+33.0	1	Planetary
C68	6729	19	02	−37.0	1	Emis. + Refl.
	6781	19	18	+06.6	2	Planetary
Blinking Star Nebula, C15	6826	19	45	+50.5	0.5	Planetary
Dumbbell Nebula, M27	6853	20	00	+22.7	6 × 9	Planetary
Crescent Nebula, C27	6888	20	12	+38.4	20 × 10	Emission
Gamma Cygni Nebula	I1318	20	16	+41.8	45 × 25	Emission
Cygnus Rift		20	40	+42		Dark
Veil Nebula West, C34	6960	20	46	+30.7	70 × 6	Emission
Veil Nebula East, C33	6992	20	56	+31.7	60 × 8	Emission
North American Nebula, C20	7000	20	59	+44.4	120 × 100	Emission
C4	7023	21	02	+68.2	18 × 18	Reflection
Egg Nebula		21	02	+36.7		Bright
Saturn Nebula, C55	7009	21	04	−11.4	4	Planetary
		21	53	+47.2	100 × 10	Dark
Cocoon Nebula, C19	I5146	21	54	+47.3	12 × 12	Emission
Helix Nebula, C63	7293	22	30	−20.8	13	Planetary
Cave Nebula, C9		22	57	+62.6	50 × 30	Emission
Bubble Nebula, C11	7635	23	21	+61.2	15 × 8	Emission
C22	7662	23	26	+42.6	0.3	Planetary

Table A4.4. Galaxies

Galaxy	NGC	RA$_{2000}$ (h	m)	Dec$_{2000}$ (')	m$_v{}^a$	Angular size (')	Type
C17	147	00	33	+48.5	9.3	8 × 13	E4
C18	185	00	39	+48.3	9.2	10 × 12	E0
	205	00	40	+41.7	8.0	10 × 17	E6
M32	221	00	43	+40.8	8.2	6 × 8	E2
Andromeda galaxy (M31)	224	00	43	+41.3	3.5	1° × 3°	Sb
C62	247	00	47	−20.8	8.9	7 × 20	S
Sculptor galaxy (C65)	253	00	48	−25.3	7.1	7 × 25	Sc
Small Magellanic Cloud		00	53	−72.8	2.3	2.5° × 5°	SB
IC 1613 (C51)		01	05	+02.1	9.3	11 × 12	Irr
Triangulum galaxy (M33)		01	34	+30.7	5.7	40 × 60	Sc
M74	628	01	37	+15.8	9.2	10 × 10	Sc
C23	891	02	23	+42.4	9.9	3 × 14	Sb
	1023	02	40	+39.1	9.5	3 × 9	E7
M77 (3C 71)	1068	02	43	0.0	8.8	6 × 7	Sb
C67	1097	02	46	−30.3	9.3	7 × 9	SBb
	1232	03	10	−20.6	9.9	7 × 8	Sc
	1291	03	17	−41.1	8.5	9 × 11	SBa
	1316	03	23	−37.2	8.9	6 × 7	S0
	1365	03	34	−36.1	9.5	6 × 10	SBb
	1399	03	39	−35.5	9.9	3 × 3	E1
	1398	03	39	−26.3	9.7	5 × 7	SBb
	1549	04	16	−55.6	9.9	3 × 4	E0
	1553	04	16	−55.8	9.5	3 × 4	S0
	1566	04	20	−54.9	9.4	6 × 8	SBb
	1808	05	08	−37.5	9.9	4 × 7	SBa
Large Magellanic Cloud		05	24	−70	0.1	9° × 11°	Irr
C7	2403	07	37	+65.6	8.4	11 × 18	Sc
	2683	08	53	+33.4	9.7	3 × 9	Sb
	2841	09	22	+51.0	9.3	4 × 8	Sb
	2903	09	32	+21.5	8.9	7 × 13	Sb
M81	3031	09	56	+69.1	6.9	14 × 26	Sb
M82 (3C 231)	3034	09	56	+69.7	8.4	5 × 11	Pec
	3077	10	03	+68.7	9.9	4 × 5	E2
Spindle galaxy (C53)	3115	10	05	−07.7	9.2	3 × 8	E6
UGC 5470		10	08	+12.3	9.8	8 × 11	E3
	3184	10	18	+41.4	9.8	7 × 7	Sc
M95	3351	10	44	+11.7	9.7	5 × 7	SBb
M96	3368	10	47	+11.8	9.2	5 × 7	Sb
M105	3379	10	48	+12.6	9.3	4 × 5	E1
	3521	11	06	0.0	8.9	5 × 10	Sb
M108	3556	11	12	+55.7	10.1	3 × 8	Sc
M65	3623	11	19	+13.1	9.3	3 × 10	Sb
M66	3627	11	20	+13.0	9.0	4 × 9	Sb
	3628	11	20	+13.6	9.5	4 × 15	Sb
M109	3992	11	58	+53.4	9.8	5 × 8	SBb

(continued overleaf)

Table A4.4 (continued)

Galaxy	NGC	RA2000 (h	m)	Dec2000 (')	$m_v{}^a$	Angular size (')	Type
	4125	12	08	+65.2	9.8	3 × 5	E5
M98	4192	12	14	+14.9	10.1	3 × 10	Sb
	4214	12	16	+36.3	9.7	6 × 8	Irr
C3	4236	12	17	+69.5	9.7	7 × 19	SB
M99	4254	12	19	+14.4	9.8	5 × 5	Sc
M106	4258	12	19	+47.3	8.3	8 × 18	Sb
M61	4303	12	22	+04.5	9.7	6 × 6	Sc
M100	4321	12	23	+15.8	9.4	6 × 7	Sc
M84	4374	12	25	+12.9	9.3	4 × 5	E1
M85	4382	12	25	+18.2	9.2	5 × 7	S0
M86	4406	12	26	+13.0	9.2	6 × 7	E3
C21	4449	12	28	+44.1	9.4	4 × 5	Irr
M49	4472	12	30	+08.0	8.4	7 × 9	E4
	4490	12	31	+41.6	9.8	3 × 6	Sc
M87	4486	12	31	+12.4	8.6	7 × 7	E1
M88	4501	12	32	+14.4	9.5	4 × 7	Sb
	4526	12	34	+07.7	9.6	2 × 7	E7
	4535	12	34	+08.2	9.8	5 × 7	Sbc
M91	4548	12	35	+14.5	10.2	4 × 5	SBb
M89	4552	12	36	+12.6	9.8	4 × 4	E0
C36	4559	12	36	+28.0	9.9	5 × 11	Sc
C38	4565	12	36	+26.0	9.6	3 × 16	Sb
M90	4569	12	37	+13.2	9.5	5 × 10	Sb
M58	4579	12	38	+11.8	9.8	4 × 5	Sb
Sombrero galaxy(M104)	4594	12	40	−11.6	8.3	4 × 9	Sb
M59	4621	12	42	+11.6	9.8	3 × 5	E3
C32	4631	12	42	+32.5	9.3	3 × 15	Sc
	4636	12	43	+02.7	9.6	5 × 6	E1
M60	4649	12	44	+11.6	8.8	6 × 7	E1
C52	4697	12	49	−05.8	9.3	4 × 6	E4
	4699	12	49	−08.7	9.6	3 × 4	Sa
	4725	12	50	+25.5	9.2	8 × 11	SBb
M94	4736	12	51	+41.1	8.2	9 × 11	Sb
	4753	12	52	−01.2	9.9	3 × 5	Pec
Black-eye galaxy (M64)	4826	12	57	+21.7	8.5	5 × 9	Sb
C29	5005	13	11	+37.1	9.8	3 × 5	Sb
M63	5055	13	16	+42.0	8.6	8 × 12	Sb
	5102	13	22	−36.6	9.7	4 × 9	S0
Cen A (C77)	5128	13	26	−43.0	7.0	15 × 18	Pec
Whirlpool galaxy (M51)	5194	13	30	+47.2	8.4	8 × 11	Sc
Part of M51	5195	13	30	+47.3	9.6	4 × 5	Pec
M83	5236	13	37	−29.9	8.2	10 × 11	Sc
Pinwheel galaxy (M101)	5457	14	03	+54.4	7.7	26 × 27	Sc
Circinus galaxy		14	13	−65.3	9.9	1 × 3	Sb
C12	6946	20	35	+60.2	8.9	10 × 11	Sc
C30	7331	22	37	+34.4	9.5	4 × 11	Sb
	7793	23	58	−32.6	9.1	7 × 9	S

^{a} This is the magnitude of all the stars in the galaxy added together, sometimes called the integrated magnitude.

The Greek Alphabet

Letter	Lower case	Upper case
Alpha	α	Α
Beta	β	Β
Gamma	γ	Γ
Delta	δ	Δ
Epsilon	ε	Ε
Zeta	ζ	Ζ
Eta	η	Η
Theta	θ	Θ
Iota	ι	Ι
Kappa	κ	Κ
Lambda	λ	Λ
Mu	μ	Μ

Letter	Lower case	Upper case
Nu	ν	Ν
Xi	ξ	Ξ
Omicron	ο	Ο
Pi	π	Π
Rho	ρ	Ρ
Sigma	σ	Σ
Tau	τ	Τ
Upsilon	υ	Υ
Phi	φ	Φ
Chi	χ	Χ
Psi	ψ	Ψ
Omega	ω	Ω

Constellations

Constellation	Abbreviation	Genitives
Andromeda	And	Andromedae
Antlia	Ant	Antliae
Apus	Aps	Apodis
Aquarius	Aqr	Aquarii
Aquila	Aql	Aquilae
Ara	Ara	Arae
Aries	Ari	Arietis
Auriga	Aur	Aurigae
Boötes	Boo	Boötis
Caelum	Cae	Caeli
Camelopardalis	Cam	Camelopardalis
Cancer	Cnc	Cancri
Canes Venatici	CVn	Canum Venaticorum
Canis Major	CMa	Canis Majoris
Canis Minor	CMi	Canis Minoris
Capricornus	Cap	Capricorni
Carina	Car	Carinae
Cassiopeia	Cas	Cassiopeiae
Centaurus	Cen	Centauri
Cepheus	Cep	Cephei
Cetus	Cet	Ceti
Chamaeleon	Cha	Chamaeleontis
Circinus	Cir	Circini
Columba	Col	Columbae
Coma Berenices	Com	Comae Berenices
Corona Australis	CrA	Coronae Australis
Corona Borealis	CrB	Coronae Borealis
Corvus	Crv	Corvi
Crater	Crt	Crateris
Crux	Cru	Crucis
Cygnus	Cyg	Cygni
Delphinus	Del	Delphini
Dorado	Dor	Doradus
Draco	Dra	Draconis

Constellation	Abbreviation	Genitives
Equuleus	Equ	Equulei
Eridanus	Eri	Eridani
Fornax	For	Fornacis
Gemini	Gem	Geminorum
Grus	Gru	Gruis
Hercules	Her	Herculis
Horologium	Hor	Horologii
Hydra	Hya	Hydrae
Hydrus	Hyi	Hydri
Indus	Ind	Indi
Lacerta	Lac	Lacertae
Leo	Leo	Leonis
Leo Minor	LMi	Leonis Minoris
Lepus	Lep	Leporis
Libra	Lib	Librae
Lupus	Lup	Lupi
Lynx	Lyn	Lyncis
Lyra	Lyr	Lyrae
Mensa	Men	Mensae
Microscopium	Mic	Microscopii
Monoceros	Mon	Monocerotis
Musca	Mus	Muscae
Norma	Nor	Normae
Octans	Oct	Octantis
Ophiuchus	Oph	Ophiuchi
Orion	Ori	Orionis
Pavo	Pav	Pavonis
Pegasus	Peg	Pegasi
Perseus	Per	Persei
Phoenix	Phe	Phoenicis
Pictor	Pic	Pictoris
Pisces	Psc	Piscium
Piscis Austrinus	PsA	Piscis Austrini

(continued overleaf)

Constellation	Abbreviation	Genitives
Puppis	Pup	Puppis
Pyxis	Pyx	Pyxidis
Reticulum	Ret	Reticuli
Sagitta	Sge	Sagittae
Sagittarius	Sgr	Sagittarii
Scorpius	Sco	Scorpii
Sculptor	Scl	Sculptoris
Scutum	Sct	Scuti
Serpens	Ser	Serpentis
Sextans	Sex	Sextantis
Taurus	Tau	Tauri

Constellation	Abbreviation	Genitives
Telescopium	Tel	Telescopii
Triangulum	Tri	Trianguli
Triangulum Australe	TrA	Trianguli Australis
Tucana	Tuc	Tucanae
Ursa Major	UMa	Ursae Majoris
Ursa Minor	UMi	Ursae Minoris
Vela	Vel	Velorum
Virgo	Vir	Virginis
Volans	Vol	Volantis
Vulpecula	Vul	Vulpeculae

Appendix 7

Useful World-Wide-Web and Internet Addresses

The pages of the Internet change daily and there are a very large number of addresses containing astronomical data. Often sites come and go in short periods of time but the URLs listed here are a selection that should remain in existence for some time. Many of them contain numerous links to other sites so you will be able quickly to find pages on topics not covered here directly. Some of them have mirror sites which may provide quicker access.

Astronomical equipment	http://Astronomy-Mall.com
Astronomical Society of the Pacific	http://maxwell.sfsu.edu/asp/asp.html
Astronomy: Good Starting Points	http://www.herts.ac.uk/lrc/subjects/natsci/astrhpg.htm
Astronomy Now	http://www.demon.co.uk/astronow
Astronomy Text Book	http://www.herts.ac.uk/astro_ub/
Comet information	http://encke.jpl.nasa.gov
Comet information	http://seds.lpl.arizona.edu/billa/tnp/comets.html
Digitised Sky Survey	http://stdatu.stsci.edu/cgi-bin/dss_form
European Southern Observatory	http://www.eso.org
HST Guide Star Catalogue	http://arch-http.hq.eso.org/cgi-bin/gsc
Hubble Space Telescope	http://www.stsci.edu
International Dark Skies Association	http://www.darksky.org
IOTA asteroid occultations	http://www.anomalies.com/iota/splash.htm
IOTA lunar occultations and eclipses	http://www.sky.net/~robinson/iotandx.htm
Jet Propulsion Laboratory	http://www.jpl.nasa.gov
Messier object images	http://seds.lpl.arizona.edu/messier/Messier.html
Optec (photometers)	http://www.optecinc.com
SBIG (CCDs)	http://www.sbig.com
Shareware and PD software	http://www.prenhall.com/~chaisson/download
Sky and Telescope	http://www.skypub.com
Solar eclipse data	http://umbra.nascom.nasa.gov.eclipse
Solar image taken today	http://www.sel.bldrdoc.govt/today.html
Solar physics	http://umbra.nascom.nasa.gov
Solar System images and data	http://bang.lanl.gov/solarsys
Space Telescope Science Institute	http://marvel.stsci.edu/
Stars	http://guinan.gsfc.nasa.gov
Stars and Galaxies	http://www.eia.brad.ac.uk/btl/
UK amateur astronomy	http://www.ukindex.co.uk/ukastro

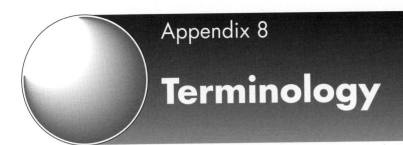

Terminology

A few technical terms that may be useful are defined below:

Aeon – a time interval of 1 000 000 000 years.

Aperture – the diameter of the objective.

Degree (°) – a measure of angle equal to 1/360 of a complete circle. It is divided into 60 minutes of arc ('), each of which is in turn divided into 60 seconds of arc (").

Diffraction – a phenomenon arising from the wave nature of light, which means that light passing an edge (of a lens, mirror, supporting strut etc.) is sent off in many directions, not just the direction that you might expect from geometrical optics. With small telescopes, diffraction is the main limit on the resolution of the instrument.

Effective focal length – the focal length which results from the combined effect of all the optical components of the instrument. In many designs of telescope, such as the Cassegrain and Schmidt–Cassegrain, the focal length of the primary mirror is quite short but the secondary mirror then increases this by a factor of 3 or 4, giving a compact design with a long focal length. The effect of Barlow lenses and telecompressors is also to extend or reduce the effective focal length of a telescope.

Eyepiece – the small lens or set of lenses used to produce an image visible to the eye from the light focused by the objective (Section 2.2).

Focal length – the distance from the centre of the lens or mirror to the point at which light from a very distant object is brought to a focus.

Focal ratio – the ratio of the focal length of a lens or mirror to its diameter (aperture).

Minutes of arc (') – see Degree.

Objective – this is the main light-gathering and focusing part of the telescope. It is either a large lens (in a refractor), or a mirror (in a reflector).

Plate scale – the linear distance along the image plane of a telescope, which corresponds to a unit angular separation of two objects. It is the scale that you will get on the negative of a photograph taken through the telescope or on the chip of a CCD detector. For an effective focal length of 2 m, the plate scale is 0.6 mm per minute of arc.

Resolution – the ability of a telescope to enable two close objects (such as stars) to be seen separately (Section 2.5).

Seconds of arc (") – see Degree.

Index

Index

186

Index

Optimum telescope for lunar work 60–61
Orbits
  elliptical 76
  Mars 69
Orbital inclinations 77
Orbital parameters 65
Orbital periods 62
Orion 2, 8, 94
Orion nebulae 98, 139, 155
Orion's belt 96, 100
Orthoscopic eyepieces 28
Parabolas 23
Parabolic mirrors 24
Parallax 62
Parameters, orbital 65
Parfocal eyepieces, set of 28
Parhelia 130
Partial eclipse 57
Partially reflecting mirrors 41
Penumbra 133
Perihelion 76, 81
Periodic comets, definition of 76
Periods
  orbital 62
  rotational 62
    different 71
  synodic 65–66, 69–70
Personal correction factor 52
Phases 68
  of Mercury 66–68
  of the Moon 55–56, 60, 63, 148
  of Venus 69
  waning 148
  waxing 148
Phobos 70
Photo-multiplier tubes 144
Photodiodes 144
Photographic photometry 144–145
Photographs 19, 30, 32, 46, 63, 80, 135
  direct 26
Photography 23, 39, 138–142, 144
  focal plane through the telescope 139
  piggyback 138–139
Photometers 145
  designs of 145
  virtual 146
Photometric variable stars 86, 88
Photometry 144–147, 152
  CCD 146
Pickering's scale 45
Piggyback photography 138–139
Pisces 15
Pixels 143
Plages 51–52
Planetarium programs 150–151
Planetary ephemeris 17
Planetary nebulae 93, 99–100, 108–113, 137
Planetary observations 19, 137
Planets 18, 65–73, 132
  lunar occultations of 149
Plate scale 183
Platforms, inclined plane equatorial-drive 41
Pleiades 89
Plössl eyepiece 28
Plough 2–3, 5–6
Pluto 65, 72–73
Pogson's equation 83
Point source 83
Point spread functions 42, 61–62
Pointers 7, 9
Polar axis 30, 32
Polaris 7, 30, 84
Pole Star 7, 30, 84
Poles, magnetic 132
Practical astronomy books 162
Precautions 49
Precession 7, 17
Primary minimum 87

Processing, choice of 141–142
Profiles, instrumental 42, 61–62
Programmes 18
  observing 21–22
  planetarium 150–151
  solar observing 52–53
Prominence spectroscope 53
Prominences 53
Protostars 99
Ptolemy 132
Pupil 27, 36
  exit 27
Purpose-designed CCD cameras 143
QSOs 126
Quadrature 150
Quadrature east 66
Quadrature west 66
Quasars 114–115, 126–128
Quasi-stellar objects 126
R CrB stars 87
R Monocerotis 97
Radar 68
Radiant of a meteor shower 131
Radiation, H-α 138
Radio telescopes 93
Rainbows 130
Ramsden eyepiece 28
Ratio, focal 183
Raw data 151
Rays, crater 59
Real image 26
Reciprocity failure 142
Recurrent novae stars 87
Red lighting 37
Redshift 114
Reflection nebulae 89, 93, 95–99, 138
Reflective coatings 22, 34–35
Reflectors 19, 21, 61
  Cassegrain 24–25
  Newtonian 22–24
Refraction 40–41
Refractor 19, 21, 29, 55, 61
Resolution 33–34, 42, 50, 61, 183
  diffraction-limited 26, 42
Retina 37
Retrograde motion 65, 132
Rhea 72
Rhodopsin 37
Rift valley 69
Right ascension 1, 7, 15, 30
Ring around the Moon 129
Ring nebula 108–109, 155
Rings 71–72
Ritchey-Chrétien reflector 24
River beds 69
Rod cells 33, 37
Rosse, Lord 117
Rotation 80
  of the asteroids 73
  of Mars 70
  of Mercury 67
  Sun, period of 52
Rotational period 62
  different 71
Rotational velocity 54
RR Lyrae variables 146
RZ Cas 146
Satellites
  Galilean 71
  Jupiter, Galilean 65
Satellite eclipses 150
Satellite galaxies 90
Saturn 65–72, 150
  meteorology 72
Saturn nebula 110–111
Scales
  Antoniadi's 46
  Pickering's 45
Schmidt–Cassegrain telescope 22, 25, 29, 183
Scintillation 26, 43–44, 149

Screens 50
Seasons 1
Secondary minimum 87
Secondary mirrors 23–24, 43
  supports for 20
Seconds of arc (") 183
Seeing 26, 38, 43–44
Seeing disk 38
Setting circles 14, 17, 65–66
Seven Sisters 89
Seyfert galaxies 114–115, 126–128
Shadow, Earth's 133
Shifts, Doppler 34, 54
Shoemaker-Levy 9 71, 79
Shooting stars 130
Short-period comets 76
Sidereal day 17
Sidereal time 17
Siderostat 153
Silicon monoxide 35
Silver coatings 35
Silver mirror test 35
Single lens reflex (SLR) cameras 139–140, 142
Sirius 8, 34, 42, 84–85, 88
61 Cygni 152
Sky background 45, 135, 137
Slide films 142
Slit spectrographs 154
Slow motion drive 30
SLR (single lens reflex) cameras 139–140, 142
Small Magellanic cloud 126
Societies
  Astronomical 157–159
  national astronomical 46
Solar diagonal 51
Solar eclipse 52, 57
  flare 52
  halo 129
  photosphere 54
  spectroscopy 54
  wind 77
Solar observing programmes 52–53
Sombrero galaxy 119
South tropical disturbance 71
Southern Cross 2
Southern pole, pointers to 13
Spacecraft 132–133
Speckle interferometry 44
Spectra
  emission nebulae 99–100
  Nova Cyg 1992 154
  planetary nebula 100
Spectral types 85
Spectrographs 113, 153
  direct vision 154
  slit 154
Spectroheliscopes 53–54
Spectroscopic analysis 93
Spectroscopic binaries 88
Spectroscopy 153–155
  solar 54
Spectrum
  absorption line 113
  emission line 113
Spherical aberration 23
Spherical primary mirror 25
Spikes 20, 60
  diffraction 42–43
Spiral arms 116
Spiral galaxies 99, 114–120, 138
  barred 114
  normal 114
Sporadic meteors 130
Spots 71, 72
Spring tides 129
Standard stars 84
Star atlas 7, 45
Star catalogues 6, 17, 18, 45, 84, 161
Star charts 17, 18
Star clusters 89–93, 126
Star diagonals 28–30

Star formation 98
Star hopping 1, 8–13, 17, 39, 44, 80
Star-shaped masks 42–43
Stars 83–92
  binary 88–89
  close double 149
  comparison 85
  double 26
    colour contrasts in 89
  double close 88
  dwarf 87
  eclipsing binary 145
  Evening 133
    images of 21
    immersed 149
  Pole 84
  R CrB 87
  recurrent novae 87
  standard 84
  T Tauri 90
  variable 46, 86–88, 144, 146
    General Catalogue of 86
    observing 86
    photometric 86
  visual double 88–89
  white dwarf 108
Starter constellation 7
Stellar magnitude, limiting 80
Stopping-down 49–50, 53
Straight wall 60
Sun 49–54, 129–130
  finding 52
  observing 49
  period of rotation of 52
Sun dog 130
Sundial 130
Sunspot 49–52
  maximum 49
  Zurich number 52
Super-massive black holes 114, 121
Superior conjunction 66
Supernovae 18, 87, 104–105
Supernovae remnants 93, 104–108, 113
Surface temperature 52
Swan bands 136
Swan nebulae 99, 102
Synodic periods 65–66, 69–70
T association 89
T Tauri star 90
T-grain film 142
T-ring 136, 139, 140, 142
Tail 77–78
  dust 77
  gas 77
Techniques, observing 36–43
Tele-extenders 139–141
Telecompressor lenses 140–141
  using 139
Telescopes
  computer-controlled 17
  designs of 19–25
  finder 38–40, 51
  focal plane photography through the 139
  guide 40
  guiding 80
  Hooker 113
  Hubble Space 103
  Keck 33
  lunar work, optimum for 60–61
  Newtonian 24
  photography with 138–142
  radio 93
  Schmidt–Cassegrain 22, 25, 29
Temperature, surface 52
Terminator 59
Terrae 58
Tests, silver mirror 35
Tethys 72
Thick atmosphere 68
Thin crescent 55, 58
Tidally locked 67

Tides 129
  neap 129
  spring 129
TIFF 151
Time
  civil 130
  equation of 130
  exposure 139, 143
Timekeeping, international civil 148
Titan 72
TLP (transient lunar phenomena) 55, 61–62
Total eclipse 55, 57
Trailed image 40
Transient lunar phenomena (TLP) 55, 61–62
Transit 67, 69, 71
Transmission grating 53
Trapezium 100–101
Tri-colour imaging 138
Trifid nebulae 99, 102
Triton 72
Turbulence 38, 43, 50
  atmospheric 34
  low-altitude 44
Turbulent zones 44
Twinkling 26, 43–44
UFOs 133
Ultra-wide-angle eyepieces 27
Umbra 133
Unaided observations 129–133
Uranus 65, 72–73
Ursa Major 2–3
  changing appearance of 5
  names of the main stars in 6
Ursa Minor (Little Bear) 8
Valley 58
  rift 69
Variable stars 46, 86–88, 144, 146
Variable-density filter 43
Variables 89
  cataclysmic 87
  extrinsic 86
  intrinsic 86
  irregular 87
  long-period 154
  Mira-type 87
Veil nebula 106–107
Velocity, rotational 54
Venus 65–72, 68–69, 133, 137
  phases of 69
Vernal equinox 15
Vignetting 85
Virgo A 122
Virgo cluster of galaxies 122
Virtual photometer 146
Vision, averted 37
Visual brightnesses 144
Visual double stars 88–89
Visual magnitudes, estimating 85
Visual purple 37
Volcano 69
Western quadrature 66
Waning phase 148
Washing, gentle 35
Waxing phase 148
Web, world-wide 181
Wedge, Herschel 51
Whirlpool Galaxy 117, 133
White dwarf star 108
Wild Duck cluster 89–90
Wind, solar 77
Wind shield 44
World-wide web 181
Zeeman effect 54
Zenith, arc near the 130
Zenithal hourly rate (ZHR) 131
ZHR (zenithal hourly rate) 131
Zodiac 77
Zodiacal light 131–132
Zone, turbulent 44
Zurich sunspot number 52